# LAGRANGIAN AND HAMILTONIAN MECHANICS

# LAGRANGIAN AND HAMILTONIAN MECHANICS

**M. G. CALKIN**
Dalhousie University

World Scientific
NEW JERSEY · LONDON · SINGAPORE · BEIJING · SHANGHAI · HONG KONG · TAIPEI · CHENNAI

*Published by*

World Scientific Publishing Co. Pte. Ltd.
5 Toh Tuck Link, Singapore 596224
*USA office:* 27 Warren Street, Suite 401-402, Hackensack, NJ 07601
*UK office:* 57 Shelton Street, Covent Garden, London WC2H 9HE

**Library of Congress Cataloging-in-Publication Data**
Calkin, M. G. (Melvin G.)
Lagrangian and Hamiltonian mechanics / M.G. Calkin.
p. cm.
Includes bibliographical references and index.
ISBN-13 978-981-02-2672-5
ISBN-10 981-02-2672-1
1. Hamiltonian systems. 2. Lagrange equations. 3. Mathematical physics.
I. Title.
QC20.7.H35 C35 1996
531'.01'51474--dc21
96203992

**British Library Cataloguing-in-Publication Data**
A catalogue record for this book is available from the British Library.

# PREFACE

This book grew out of notes for a one-semester course in classical mechanics which I have taught for many years to senior and, more recently, junior physics students at Dalhousie University. These students have normally taken one semester in mechanics at the introductory level and one semester at the second year level; my course is their final exposure to mechanics as undergraduates. The original aim of the course was to introduce the student, in the more familiar setting of classical mechanics, to those ideas and terms which he or she would later encounter in modified form in quantum mechanics: Lagrangian, action, Hamiltonian, Poisson brackets, canonical transformations, etc. In recent years, with the resurgence of interest in mechanics, especially non-linear dynamics, the emphasis of the course has shifted somewhat to include these contemporary developments as well.

The resulting book now contains more material than can be covered comfortably in a one-semester course. Experienced instructors can judge for themselves what material their students should omit on first reading; this depends on the level of the course and the emphasis the instructor wishes to place on the subject. While the book was written primarily as a fourth year text, the range of difficulty is quite broad. It can thus be used in third year, as I now do, simply by choosing the topics appropriately. Doing exercises forms an important part of any learning experience. The ones here have been chosen not only to exercise the students' minds, but also to provide additional examples and occasionally to introduce new topics.

Classical mechanics is a mature subject, and many others have written on it from their personal perspectives. I have indicated by footnotes throughout the book some of these other discussions which I have enjoyed studying and to which a student may refer to get a different point of view or a fuller treatment of some topic.

I wish to thank my students for encouraging me to organize my perspective of mechanics in this more formal way. My friend and colleague, David Kiang, has been most helpful. He read an early version of the manuscript and made many valuable suggestions for clarifying and improving the presentation and has been a continual source of advice on all aspects of the work. The support of my family is much appreciated. In particular, I thank my daughter Catherine for reading and editing my English; any smoothness in the writing is the result of her efforts; the roughness which remains, I take full responsibility for. My wife Patricia, good bow-paddler that she is, has helped me to avoid the rocks and shoals in this project, as in all aspects of our canoe trip together. I thank her from the bottom of my heart.

Melvin G. Calkin
Halifax, Nova Scotia
November, 1995

# PREFACE

# CONTENTS

# CHAPTER I

# NEWTON'S LAWS

The aim of classical mechanics is to describe and predict the motion of bodies and systems of bodies which are subject to various interactions. Newton's three laws of motion form the basis of this description, so let us begin by reviewing these.

## Newton's laws[1]

**Newton's first law** deals with non-interacting bodies. It says that the **velocity** of an isolated body, one removed from the influence of other bodies, is constant. This law defines a set of preferred coordinate frames, **inertial frames,** as frames in which Newton's first law holds. Given an inertial frame, we can obtain others by translating the original in space and time, by rotating the original through some angle about some axis, or by giving the original frame a uniform velocity. Unless stated otherwise, we always refer motion to an inertial frame. To a first approximation a coordinate frame attached to the earth is an inertial frame. However, various physical phenomena such as the behavior of a Foucault pendulum, the flight of a ballistic missile, atmospheric and ocean currents, indicate its true non-inertial nature. Better approximations to inertial frames are frames attached to the sun or, even better, to the "fixed stars."

Newton's second and third laws deal with the effects of interaction between bodies on their motion. Interactions cause the velocities of the bodies to change; the bodies undergo **acceleration.** Let us consider two otherwise isolated interacting small bodies, "particles." We find that the accelerations $\mathbf{a}_1$ and $\mathbf{a}_2$ of the particles are oppositely directed, and that their magnitudes are related by

$$a_1/a_2 = k_{12},$$

where the ratio $k_{12}$ is *independent* of the nature of the interaction between the particles. It does not depend on whether the interaction arises because the particles are in contact with one another, or because they are connected by a string or by a spring, or because they interact gravitationally, etc. The ratio $k_{12}$ is thus a quantity which we can associate with the pair of particles themselves, as opposed to the particular interaction they happen to be undergoing. Further, if we consider three particles, a pair at a time, we find that the three acceleration ratios are not independent but are related by

$$k_{12}k_{23}k_{31} = 1.$$

This, together with $k_{12} = 1/k_{21}$, shows that the acceleration ratio can be written

$$k_{12} = m_2/m_1,$$

---

[1]Ernst Mach, *The Science of Mechanics,* (The Open Court Publishing Co., Chicago, 1893), trans. Thomas J. McCormack.

where $m_1$ ($m_2$) is a property of, something associated with, particle 1 (particle 2) alone. The quantity $m_1$ ($m_2$) is called the **inertial mass** of the particle. Putting these facts together, we see that the accelerations of two interacting particles are related by

$$m_1\mathbf{a}_1 = -m_2\mathbf{a}_2.$$

We describe the interaction by saying that particle 2 exerts a **force** $\mathbf{F}_{2\text{on}1}$ on particle 1, and particle 1 exerts a force $\mathbf{F}_{1\text{on}2}$ on particle 2, such that

$$m_1\mathbf{a}_1 = \mathbf{F}_{2\text{on}1} \quad \text{and} \quad m_2\mathbf{a}_2 = \mathbf{F}_{1\text{on}2}$$

with

$$\mathbf{F}_{2\text{on}1} = -\mathbf{F}_{1\text{on}2}.$$

This last equation is **Newton's third law**: the force which particle 2 exerts on particle 1 is equal and opposite to the force which particle 1 exerts on particle 2. It is another way of stating our conclusions about the acceleration ratio of two interacting particles.

If we now consider three interacting particles, we find that the acceleration of any one of them, say particle 1, is the vector sum of the acceleration of particle 1 due to particle 2 alone and the acceleration of particle 1 due to particle 3 alone (Fig. 1.01(a)), and thus

$$\begin{aligned} m_1\mathbf{a}_1 &= \mathbf{F}_{2\text{on}1} + \mathbf{F}_{3\text{on}1} \\ &= \mathbf{F}_{\text{total on}1}. \end{aligned}$$

This is **Newton's second law**: the acceleration of a particle is directly proportional to the total force acting on it (obtained by adding vectorially all the individual forces) and is inversely proportional to the mass of the particle. This law should be understood in the following way: we are meant to describe the interaction of our chosen particle with other particles by specifying the force acting on it in terms of the locations and velocities of all the particles. The form this takes depends, of course, on the nature of the interactions.

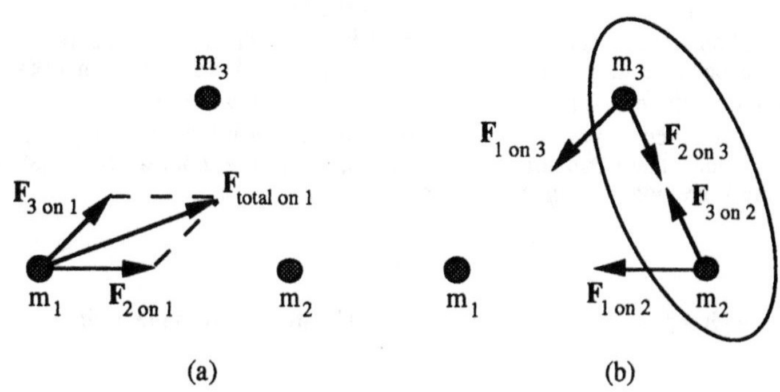


Fig. 1.01. Forces on three interacting particles

Similar equations apply to particles 2 and 3 (Fig. 1.01(b)),

$$m_2\mathbf{a}_2 = \mathbf{F}_{3\text{on}2} + \mathbf{F}_{1\text{on}2}$$
$$m_3\mathbf{a}_3 = \mathbf{F}_{1\text{on}3} + \mathbf{F}_{2\text{on}3}\,.$$

Now suppose that the forces $\mathbf{F}_{3\text{on}2}$ and $\mathbf{F}_{2\text{on}3}$ are such that particles 2 and 3 are bound together to form a single particle and move with a common acceleration

$$\mathbf{a}_2 = \mathbf{a}_3 = \mathbf{a}_{23}.$$

Then, adding the above two equations, we find

$$(m_2 + m_3)\mathbf{a}_{23} = \mathbf{F}_{1\text{on}2} + \mathbf{F}_{1\text{on}3},$$

the internal forces $\mathbf{F}_{3\text{on}2}$ and $\mathbf{F}_{2\text{on}3}$ canceling because of Newton's third law. The bound combination thus behaves as a single particle with mass

$$m_{23} = m_2 + m_3;$$

mass is additive. Further, the force acting on the bound combination can be taken to be the total *external* force; the internal forces which hold the combination together need not be taken into account.

To see how to apply these laws, we begin with some simple examples, most of which will already be familiar to you.

## Free fall

For our first example, consider the motion of a body of mass m dropped near the surface of the earth. The only force acting on the body (ignoring air friction) is the downward **gravitational force** mg; here $g \approx 9.8\,\text{m/s}^2$ is the approximately constant **gravitational field** due to the earth. Newton's second law gives

$$ma = mg,$$

so, canceling the m, we see that the body falls with constant acceleration g. Indeed, since g is body-independent, all bodies fall with the *same* constant acceleration.[2] The equation of motion for the falling body takes the form (measuring x downwards)

$$\frac{d^2x}{dt^2} = g.$$

---

[2]For the consequences, see A. Einstein in H. A. Lorentz, A. Einstein, H. Minkowski, and H. Weyl, *The Principle of Relativity* (Dover Publications, New York, NY, 1923), trans. W. Perrett and G. B. Jeffery, p. 99.

Integrating, we find that the velocity is given by

$$\frac{dx}{dt} = v_0 + gt = v(t)$$

where $v_0$ is the initial velocity. Integrating once again, we find that the position of the body is given by

$$x(t) = x_0 + v_0 t + \tfrac{1}{2} g t^2$$

where $x_0$ is the initial position. The position is thus determined as a function of the time; the expression also contains two constants of integration, adjustable parameters, which are set by the initial conditions, the initial position and velocity.

## Simple harmonic oscillator

In the above example the integration could be done immediately since we knew the time dependence of the force: it was constant. Usually, however, the force is not known *a priori* as a function of time. Rather, it is known as a function of position. Take, for example, a mass m attached to a spring (Fig. 1.02).

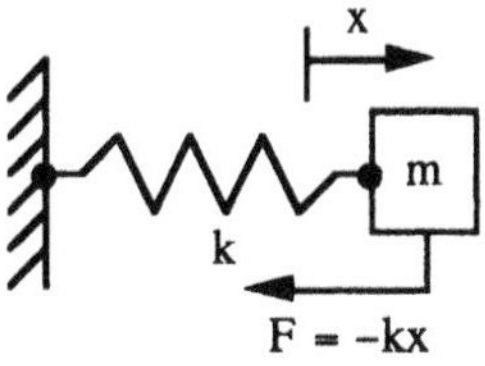


Fig. 1.02. Mass attached to a spring

It is found (**Hooke's law**) that the force F which the spring exerts on the mass is proportional to the amount x the spring is stretched and is directed opposite the stretch,

$$F = -kx.$$

The proportionality constant k, a measure of the strength of the spring, is called the **spring constant**. Newton's second law then gives

$$ma = -kx.$$

We cannot integrate this directly as in the free fall case. However, if we multiply the equation by the velocity v, the left-hand side becomes

$$mav = m\frac{dv}{dt}v = \frac{d}{dt}(\tfrac{1}{2}mv^2),$$

and the right-hand side becomes

$$-kxv = -kx\frac{dx}{dt} = -\frac{d}{dt}(\tfrac{1}{2}kx^2).$$

Thus the motion is such that

$$E = \tfrac{1}{2}mv^2 + \tfrac{1}{2}kx^2$$

is constant in time. This quantity is called the **total energy** and is the sum of the **kinetic energy** $T = \frac{1}{2}mv^2$ and the **potential energy** $V = \frac{1}{2}kx^2$.

The energy equation can be rearranged to obtain

$$\int_{x_0}^{x} \frac{dx}{\sqrt{\frac{2E}{m} - \frac{k}{m}x^2}} = \int_0^t dt = t.$$

To do the x-integration, we set

$$x = \sqrt{\frac{2E}{k}}\sin\phi \qquad dx = \sqrt{\frac{2E}{k}}\cos\phi\, d\phi$$

where $\phi$ is a new variable, the **phase**. The left-hand side then becomes

$$\sqrt{\frac{m}{k}}\int_{\phi_0}^{\phi} \frac{\cos\phi}{\sqrt{1-\sin^2\phi}}d\phi = \sqrt{\frac{m}{k}}\int_{\phi_0}^{\phi} d\phi = \sqrt{\frac{m}{k}}(\phi - \phi_0),$$

so that the phase is a linear function of time,

$$\phi = \phi_0 + \omega t.$$

The rate at which the phase increases with time, $\omega = \sqrt{k/m}$, is called the **angular frequency**. The position of the mass as a function of time is thus given by

$$x = A\sin(\omega t + \phi_0)$$

where $A = \sqrt{2E/k}$ is called the **amplitude** of the motion. The mass oscillates back and forth between $x = +A$ and $x = -A$, going through a complete cycle in a time $\tau = 2\pi/\omega$, the **period** of the motion. This very important motion is called **simple harmonic motion**. Its importance derives from the fact that, except in unusual circumstances, motion near any stable equilibrium point is simple harmonic motion.

The approach which we have used here to discuss the simple harmonic oscillator can be applied to any one-dimensional conservative system, for which $F = -dV/dx$; all we have to do is to replace $\frac{1}{2}kx^2$ by the appropriate potential energy $V(x)$. Such systems are thus in principle always integrable.

## Central force

When we move from one-dimensional problems to three-dimensional problems, the degree of complexity increases enormously. Indeed, most three-dimensional problems cannot be integrated analytically. There is, however, a class of problems which can still be handled moderately easily, namely the motion of a particle acted on by a force $\mathbf{F} = F\hat{\mathbf{r}}$ which is always directed towards (or away from) a fixed point, the **force center**. This is the so-called **central force problem** (Fig. 1.03(a)). For such problems the **torque** $\mathbf{r} \times \mathbf{F}$ on the particle about the force center is zero, and the **angular momentum** $\mathbf{L} = \mathbf{r} \times (m\mathbf{v})$ is constant. The motion thus lies in a plane $\mathbf{L} \cdot \mathbf{r} = 0$ which is perpendicular to $\mathbf{L}$ and which passes through the force center, the **orbital plane**. Further, the motion is such that the magnitude L of the angular momentum about the force center is constant.

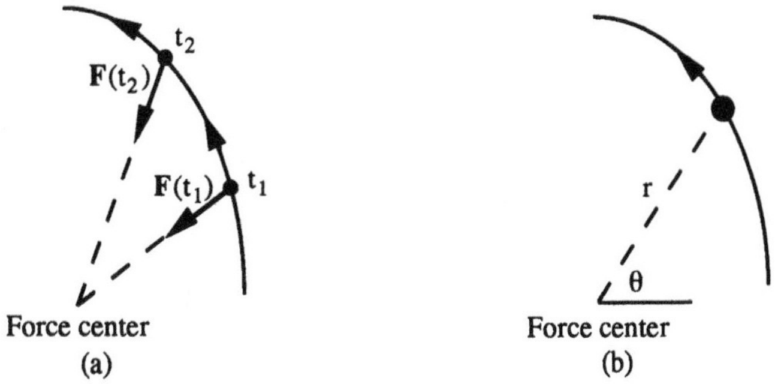


Fig. 1.03. (a) Central force, (b) Polar coordinates

It is convenient to introduce polar coordinates r and θ in the orbital plane, with the force center the pole (Fig. 1.03(b)). The position of the particle is then given by $\mathbf{r} = r\hat{\mathbf{r}}$, the velocity by

$$\begin{aligned} \mathbf{v} &= \frac{d\mathbf{r}}{dt} = \frac{dr}{dt}\hat{\mathbf{r}} + r\frac{d\hat{\mathbf{r}}}{dt} \\ &= \dot{r}\hat{\mathbf{r}} + r\dot{\theta}\hat{\boldsymbol{\theta}} \, , \end{aligned}$$

and the acceleration by

$$\mathbf{a} = \frac{d\mathbf{v}}{dt} = \ddot{r}\hat{\mathbf{r}} + \dot{r}\frac{d\hat{\mathbf{r}}}{dt} + \dot{r}\dot{\theta}\hat{\boldsymbol{\theta}} + r\ddot{\theta}\hat{\boldsymbol{\theta}} + r\dot{\theta}\frac{d\hat{\boldsymbol{\theta}}}{dt}$$

$$= (\ddot{r} - r\dot{\theta}^2)\hat{\mathbf{r}} + (r\ddot{\theta} + 2\dot{r}\dot{\theta})\hat{\boldsymbol{\theta}} \ .$$

In deriving these we have used the results

$$\frac{d\hat{\mathbf{r}}}{dt} = \dot{\theta}\hat{\boldsymbol{\theta}} \quad \text{and} \quad \frac{d\hat{\boldsymbol{\theta}}}{dt} = -\dot{\theta}\hat{\mathbf{r}}$$

which follow readily from Fig. 1.04.[3]

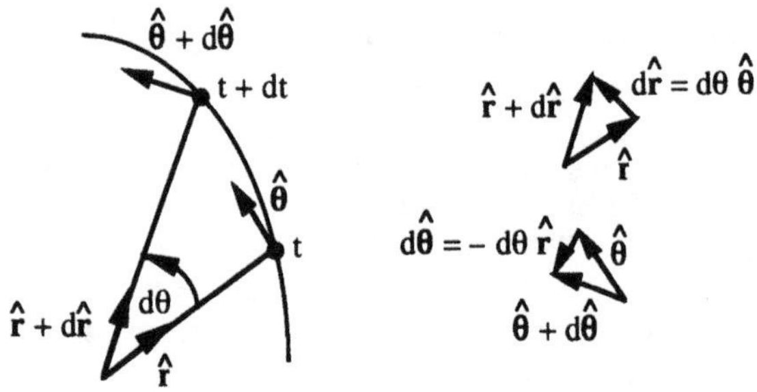


Fig. 1.04. Changes in the unit vectors

Newton's second law then gives the equations of motion

$$m(\ddot{r} - r\dot{\theta}^2) = F$$

$$m(r\ddot{\theta} + 2\dot{r}\dot{\theta}) = 0 \ .$$

The second of these can be written

$$\frac{1}{r}\frac{d}{dt}(mr^2\dot{\theta}) = 0$$

which shows that $L = mr^2\dot{\theta}$ is constant in time. The quantity L = (distance r from origin) × (component $mr\dot{\theta}$ of $m\mathbf{v}$ perpendicular to $\mathbf{r}$) is the magnitude of the angular momentum, so this simply confirms what we already know. Its content can be expressed in a rather picturesque way.

[3]Alternate derivations can be found in Daniel Kleppner and Robert J. Kolenkow, *An Introduction to Mechanics*, (McGraw-Hill Book Company, New York, NY, 1973), pp. 27-38.

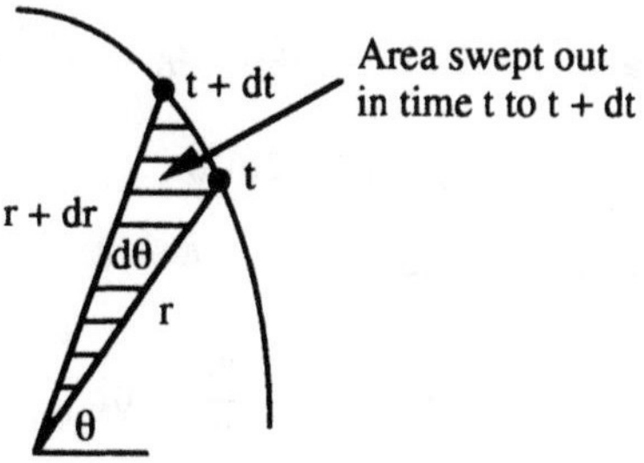


Fig. 1.05. Area swept out by radius vector

From Fig. 1.05, the radius vector sweeps out an area $dA = \frac{1}{2}(r)(r\dot{\theta}dt)$ in a time $dt$, so the rate at which it sweeps out area is $dA/dt = \frac{1}{2}r^2\dot{\theta} = L/2m$. This, as we have seen, is constant. Thus the particle moves along its orbit in such a way that the radius vector sweeps out equal areas in equal times. Applied to the solar system, this is known as **Kepler's second law** of planetary motion. We see, however, that it holds for any central force, not just for the gravitational force.

The fact that L is constant can be used to eliminate $\dot{\theta} = L/mr^2$ from the radial equation of motion. If, further, the central force is conservative, so that $F = -dV(r)/dr$ where V is the potential energy, we can write

$$m\ddot{r} = \frac{L^2}{mr^3} - \frac{dV}{dr} = -\frac{d}{dr}\left(\frac{L^2}{2mr^2} + V(r)\right).$$

The radial motion is thus the same as one-dimensional motion (but with $r > 0$) in an **effective potential**

$$V_{eff}(r) = \frac{L^2}{2mr^2} + V(r).$$

In particular, the total energy

$$E = \tfrac{1}{2}m|v|^2 + V(r) = \tfrac{1}{2}m(\dot{r}^2 + r^2\dot{\theta}^2) + V(r)$$
$$= \tfrac{1}{2}m\left(\dot{r}^2 + \frac{L^2}{m^2r^2}\right) + V(r) = \tfrac{1}{2}m\dot{r}^2 + V_{eff}(r)$$

is constant in time. This can be rearranged and integrated to obtain r as a function of t and the parameters E, L, and initial radius $r_0$. The result can then be substituted into $\dot{\theta} = L/mr^2$, and this equation integrated to obtain θ as a function of t and the parameters E, L, $r_0$, and initial angle $\theta_0$. These four parameters, together with the two required to specify the orbital plane, specify the orbit ($r_0$ and $\theta_0$ give the initial position in the orbital plane, and E and L then give the magnitude and direction of the initial velocity).

Before embarking on the integrations, either analytically or numerically, it is worthwhile to get an overview of the general behavior to be expected, to obtain qualitative pictures of the possible orbits and the ranges of the parameters over which they occur. These can be obtained by examining a sketch of $V_{eff}(r)$ versus r. We illustrate this idea by applying it to the important central force, gravitation.

## Gravitational force: qualitative

The **gravitational force** between two bodies, which are small compared to the distance r between them, is given by

$$F = -\frac{k}{r^2} = -\frac{dV}{dr}, \quad \text{where} \quad V = -\frac{k}{r}$$

is the **gravitational potential**. The constant k equals GmM where m and M are the masses of the bodies and G is the gravitational constant. The same expression may be used for the electrostatic force, the **Coulomb force**, between two electrically charged bodies, provided we set $k = -q_1q_2$ where $q_1$ and $q_2$ are the electric charges on the bodies. If one of the bodies is much lighter than the other, say $m << M$, as is the case for the planets compared with the sun, or artificial satellites with the earth, or an electron with a nucleus, we can regard the heavy body as providing an approximately fixed force center about which the lighter body orbits. The effective potential is then

$$V_{eff} = \frac{L^2}{2mr^2} - \frac{k}{r}$$

and is shown in Fig. 1.06.

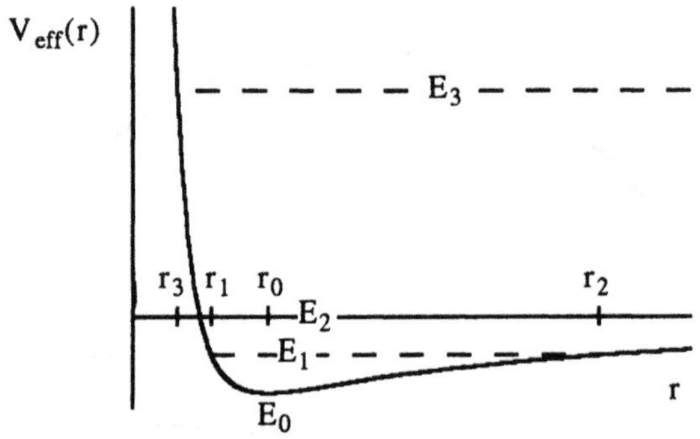


Fig. 1.06. Effective potential for the gravitational force

We have seen that the constant total energy E of an orbiting body is the sum of the kinetic energy $\frac{1}{2}m\dot{r}^2$ due to its radial motion and the effective potential $V_{eff}$. Since the kinetic energy is non-negative, the total energy must be greater than or equal to the energy $E_0$ at the bottom of the $V_{eff}$ potential well (Fig. 1.06). If the energy is $E_0$, then the radius is fixed at $r_0$ and the orbit is a circle. If the energy is $E_1$ with $E_0 < E_1 < 0$, the radial motion is like that of a particle in a one-dimensional potential well, the orbit radius oscillating back and forth between an inner turning radius $r_1$ and an outer turning radius $r_2$, with $E = V_{eff}$ and $\dot{r} = 0$ at $r_1$ and $r_2$. All the while the angle θ is increasing. We see that the orbit then looks qualitatively like one of those in Fig. 1.07.

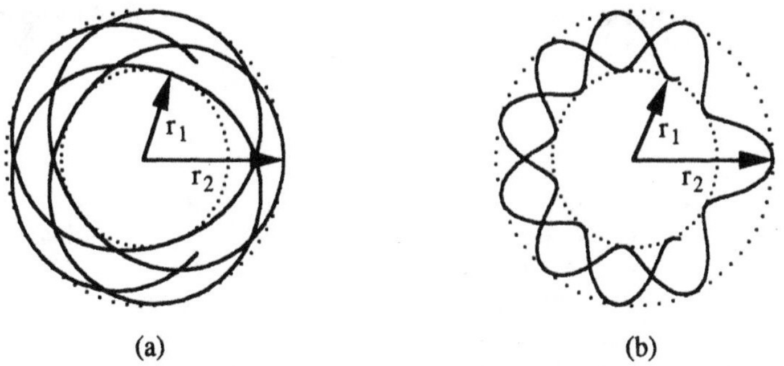


Fig. 1.07. Qualitative shape of bound orbit

Since for gravity the force is attractive rather than repulsive at the inner turning radius, the contact there must look like Fig. 1.07(a) rather than like (b). Detailed calculations to follow show that the orbit for this case is in fact an ellipse. If the energy is $E_2 = 0$ or $E_3 > 0$, there is an inner turning radius but no outer turning radius. The orbiting body comes in from infinity, reaches a minimum radius $r_3$, and moves out again to infinity. The orbit looks qualitatively like that in Fig. 1.08.

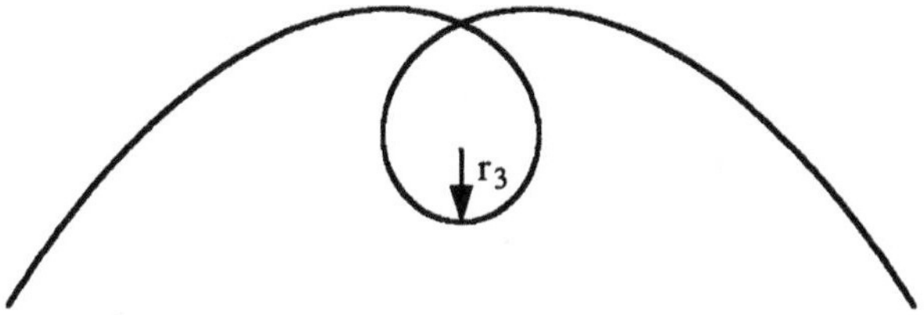


Fig. 1.08. Qualitative shape of scattering orbit

Detailed calculations to follow show that for the gravitational force the orbit is in fact a parabola for energy $E = 0$ and a hyperbola for energy $E > 0$.

The ideas used above to discuss the qualitative shape of the orbits in a gravitational potential can also be applied to an arbitrary central potential. In addition to the situations already considered for gravity, we may then encounter attractive potentials $V(r)$ which blow up faster than $-1/r^2$ as $r \to 0$. The effective potential $V_{eff}$ then tends to $-\infty$ rather than $+\infty$ as $r \to 0$ and, depending on the energy, there may be no inner turning radius. For example, an orbit with energy E shown in Fig. 1.09(a) spirals in to the force center as in Fig. 1.09(b).

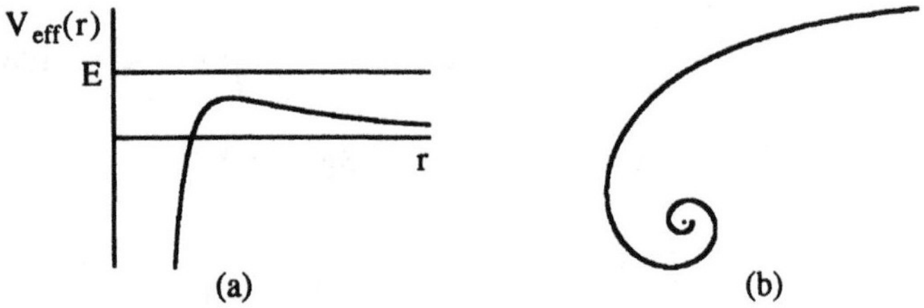


Fig. 1.09. (a) A possible effective potential, (b) Qualitative shape of capture orbit

## Gravitational force: quantitative

Let us now return to the gravitational force and consider the detailed integration. The energy and angular momentum equations

$$E = \frac{1}{2}m\dot{r}^2 + \frac{L^2}{2mr^2} - \frac{k}{r} \quad \text{and} \quad L = mr^2\dot{\theta}$$

lead, on integration, to r and $\theta$ as functions of time. Rather than doing these integrations immediately, however, it is more useful first to obtain r as a function of $\theta$; that is, to obtain the equation which describes the shape of the orbit. To do this, we set

$$\dot{r} = \frac{dr}{d\theta}\dot{\theta} = \frac{dr}{d\theta}\frac{L}{mr^2}$$

in the energy equation and hence, on rearranging, write

$$\int_{r_0}^{r} \frac{dr}{r^2\sqrt{\frac{2mE}{L^2} - \frac{1}{r^2} + \frac{2mk}{L^2 r}}} = \int_{\theta_0}^{\theta} d\theta = \theta - \theta_0.$$

The r-integration is performed by setting first $u = 1/r$, $du = -dr/r^2$ to give

$$-\int_{u_0}^{u} \frac{du}{\sqrt{\frac{2mE}{L^2}+\frac{2mk}{L^2}u-u^2}} = -\int_{u_0}^{u} \frac{du}{\sqrt{\frac{m^2k^2}{L^4}\left(1+\frac{2L^2E}{mk^2}\right)-\left(u-\frac{mk}{L^2}\right)^2}}.$$

Then set

$$u-\frac{mk}{L^2}=\frac{mk}{L^2}\sqrt{1+\frac{2L^2E}{mk^2}}\cos\alpha \quad \text{and} \quad du=-\frac{mk}{L^2}\sqrt{1+\frac{2L^2E}{mk^2}}\sin\alpha\, d\alpha.$$

Integration gives $\alpha = \theta - \theta_0$, which leads to the orbit equation

$$u=\frac{1}{r}=\frac{mk}{L^2}\left[1+\sqrt{1+\frac{2L^2E}{mk^2}}\cos(\theta-\theta_0)\right].$$

This has the form of a **conic section**

$$\frac{p}{r}=1+e\cos(\theta-\theta_0)$$

with **semi-latus-rectum** $p=\frac{L^2}{mk}$ and **eccentricity** $e=\sqrt{1+\frac{2L^2E}{mk^2}}$. We have chosen the constant of integration so that $\theta=\theta_0$ is the direction of **pericenter**, the point on the orbit closest to the force center. The angle $\alpha=\theta-\theta_0$ from pericenter is called the **true anomaly**. We can show that:

for $E=E_0=-\frac{mk^2}{2L^2}$, $e=0$ and the orbit is a **circle,**
for $E_0<E<0$, $0<e<1$ and the orbit is an **ellipse,**
for $E=0$, $e=1$ and the orbit is a **parabola,**
and for $0<E$, $1<e$ and the orbit is a **hyperbola.**

Let us first consider the bound orbits $E<0$, those for which the particle is confined to a finite region of space. We show that the above equation with $0<e<1$ represents an ellipse (clearly, the special case $e=0$ represents a circle). As the old geometry books put it, an ellipse is the locus of all points for which the sum of the distances to two fixed points is a constant. The two fixed points are called the **foci** of the ellipse. The sum of the distances is the **major axis** of the ellipse. We denote it by 2a, so a is the semi-major axis. The ratio of the distance between the foci to the major axis is the **eccentricity** e of the ellipse.

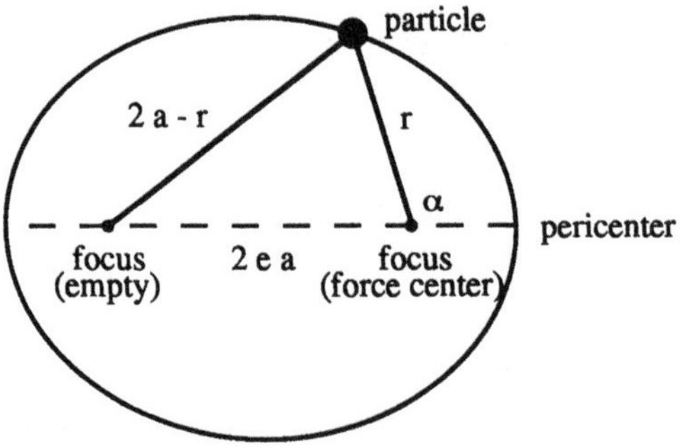


Fig. 1.10. Elliptic orbit

If we apply the trigonometric cosine law to the triangle in Fig. 1.10, we find

$$(2a - r)^2 = r^2 + 4e^2a^2 + 4ear\cos\alpha,$$

which, on rearranging, gives the polar form of the equation of an ellipse, as above, with semi-latus-rectum $p = a(1 - e^2)$. Applied to the solar system, this is **Kepler's first law** of planetary motion: the planets travel around the sun on an elliptical orbit with the sun at one focus.

The semi-major axis of the ellipse can be expressed in terms of the energy and angular momentum,

$$p = \frac{L^2}{mk} = a\frac{2L^2|E|}{mk^2} \quad \text{so} \quad a = \frac{k}{2|E|} \quad \text{or} \quad E = -\frac{k}{2a}.$$

The semi-major axis depends only on the energy, not on the angular momentum; alternatively, the energy depends only on the semi-major axis, not on the eccentricity.

Let us now turn our attention to the time dependence of the variables. According to Kepler's second law, the rate at which the radius vector sweeps out area is

$$\frac{dA}{dt} = \frac{L}{2m} = \frac{1}{2}\sqrt{\frac{ka}{m}(1 - e^2)}.$$

The time required to complete one orbit, the **period** $\tau$, is the time to sweep out the complete area

$$A = \pi ab = \pi a^2\sqrt{1 - e^2}$$

enclosed by the ellipse (here $b = a\sqrt{1 - e^2}$ is the semi-minor axis of the ellipse). It is given by

$$\frac{\pi a^2\sqrt{1-e^2}}{\tau} = \frac{1}{2}\sqrt{\frac{ka}{m}(1-e^2)}\,,$$

which, on simplifying, yields

$$\tau = 2\pi\sqrt{\frac{m}{k}}\,a^{3/2}.$$

For the family of planets orbiting the sun, $k/m = GM$ where M is the mass of the sun. Thus, the period of a planet is proportional to the $3/2$ power of the semi-major axis of the planet's orbit; it does not depend on the mass of the planet or the eccentricity of the orbit. This is **Kepler's third law** of planetary motion.[4] See Fig. 1.11 and note that the slope of the log-log plot is $3/2$.

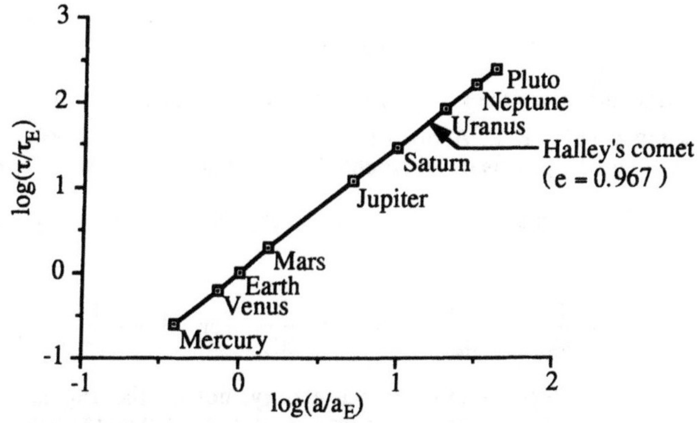


Fig. 1.11. Kepler's third law

Kepler's second law, plus some geometry, can also give the way the particle moves around the orbit as a function of time.[5] However, for modern readers more familiar with calculus than with geometry, it is probably easier to obtain this by integrating the equation for radial motion

$$\frac{1}{2}m\dot{r}^2 + \frac{L^2}{2mr^2} - \frac{k}{r} = E.$$

[4]One sometimes sees a "corrected" version in which GM is replaced by G(M+m). This comes from treating the situation as a two-body problem involving the sun and the planet under consideration. But the solar system is a *many body* problem, and the solar motions which lead to "corrections" of this type are the result of the interaction of the sun not only with the planet under consideration but with all the planets. A proper treatment would have to take all these into account.

[5]See, for example, Forest Ray Moulton, *An Introduction to Celestial Mechanics*, (Macmillan, New York, NY, 1902, 1914, 1923), 2nd ed., p.159.

This can be rearranged in the form

$$\int_{r_0}^{r} \frac{dr}{\sqrt{E + \frac{k}{r} - \frac{L^2}{2mr^2}}} = \sqrt{\frac{2}{m}} \int_{t_0}^{t} dt = \sqrt{\frac{2}{m}}(t - t_0).$$

If the energy E and angular momentum L are expressed in terms of the semi-major axis a and eccentricity e, the left-hand side becomes

$$\sqrt{\frac{2a}{k}} \int_{a(1-e)}^{r} \frac{rdr}{\sqrt{a^2e^2 - (r-a)^2}}.$$

Here we have also chosen $r_0$ as the pericenter radius $a(1-e)$; $t_0$ is then the time of passage of pericenter. The integration is readily performed by setting

$$r - a = -ea\cos\psi \quad \text{and} \quad dr = ea\sin\psi\, d\psi$$

where $\psi$ is a new variable called the **eccentric anomaly**, whose geometric significance can be seen from Fig. 1.12.

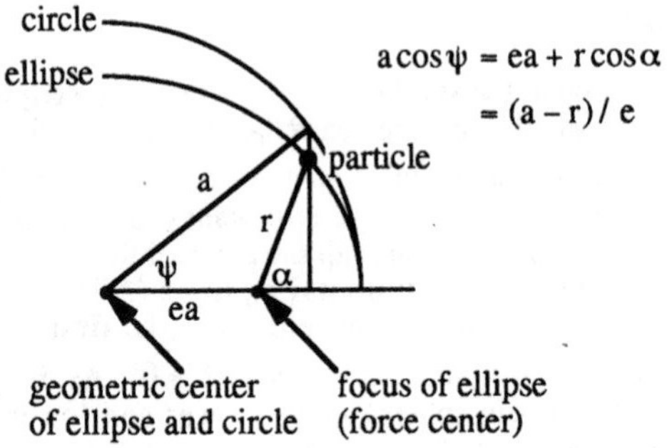


Fig. 1.12. Eccentric anomaly $\psi$

The integration then becomes

$$\sqrt{\frac{2a}{k}}\, a \int_0^{\psi} (1 - e\cos\psi)\, d\psi = \sqrt{\frac{2a}{k}}\, a(\psi - e\sin\psi),$$

and we find

$$\frac{r}{a} = 1 - e\cos\psi$$

with

$$\psi - e\sin\psi = \frac{2\pi}{\tau}(t - t_0)$$

where $\tau$ is the period of the motion. This last equation is known as **Kepler's equation**; it is a relation between the eccentric anomaly $\psi$ and the time t, or the so-called **mean anomaly** $(2\pi/\tau)(t - t_0)$. Finally, the relation between the eccentric anomaly $\psi$ and the true anomaly $(\theta - \theta_0)$ can be found by eliminating r/a between the above r-$\psi$ equation and the orbit equation, to obtain

$$1 - e\cos\psi = \frac{1 - e^2}{1 + e\cos(\theta - \theta_0)}.$$

In principle, the determination of the motion is now complete. If, however, we want r and $\theta$ in terms of the time, as is often the case, we must first invert Kepler's equation to get $\psi$ in terms of the time. This is in general difficult. In cases in which the eccentricity e of the orbit is small, however, expansions in powers of e are adequate. We use this approach in the next section to determine the parameters of earth's orbit.

## Parameters of earth's orbit

The earth's orbit around the sun lies in the plane of the **ecliptic**, which is marked by the apparent path of the sun through the constellations of the zodiac over the course of a year. The plane of the earth's equator makes an angle of approximately 23° with the plane of the ecliptic, and the intersection of these two planes gives a direction in space. In September of each year the sun passes through the plane of the earth's equator going from north to south. This is known as the **autumnal equinox** (AE). The direction from sun to earth at this time provides a convenient reference (the **first point of Aries**) from which to measure the angle $\theta$, so at autumnal equinox $\theta = 0$. As the year progresses, the midday sun (in the northern hemisphere) moves lower and lower in the sky, until at $\theta = \pi/2$, the **winter solstice** (WS) in December, it reaches its lowest point. The sun then moves higher in the sky and at $\theta = \pi$, the **vernal equinox** (VE) in March, the sun again passes through the plane of the earth's equator, this time going from south to north. The first point of Aries is thus also the direction from earth to sun at the vernal equinox. The sun continues to move higher in the sky, and at $\theta = 3\pi/2$, the **summer solstice** (SS) in June, the midday sun reaches its highest point. See Fig. 1.13.

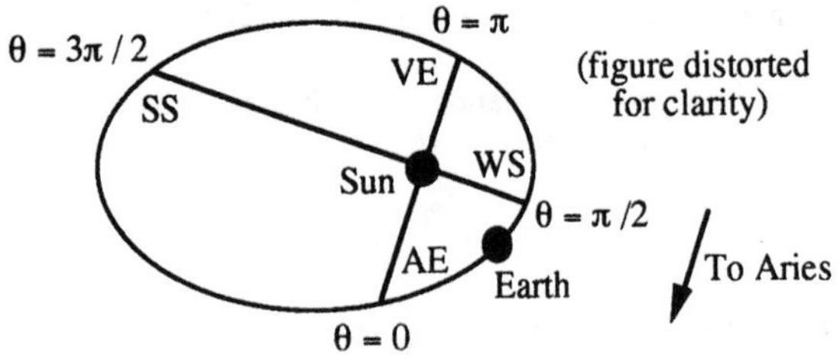


Fig. 1.13. Earth's orbit

Table I. Equinoxes, solstices, and seasons for 1994 - 1995

| | Equinoxes and solstices | | Seasons | | |
|---|---|---|---|---|---|
| | Date (see 1) | Day (see 2) | (see 3) | (see 4) | |
| AE (94) | 23 Sept. 01:19 | 266.0549 | | | |
| | | | 89.8361 | 0.245961 | Autumn |
| WS (94) | 21 Dec. 21:23 | 355.8910 | | | |
| | | | 88.9938 | 0.243654 | Winter |
| VE (95) | 20 Mar. 21:14 | 444.8847 | | | |
| | | | 92.7639 | 0.253977 | Spring |
| SS (95) | 21 June 15:34 | 537.6486 | | | |
| | | | 93.6521 | 0.256408 | Summer |
| AE (95) | 23 Sept. 07:13 | 631.3007 | | | |

1. Times in EST
2. 1 Jan. 1994 0:00 h = 1.0000
3. In days
4. In fractions of the year [AE (94) to AE (95)] = 365.2458 days

We wish to use the "observed" times of these seasonal events[6] to find the parameters of earth's orbit: the eccentricity e which determines the shape of the ellipse, the angle $\theta_0$ of perihelion which determines the orientation of the ellipse in the plane of the ecliptic, and the time $t_0$ of passage of perihelion. The last equation of the previous section gives the relation between the eccentric anomaly $\psi$ and the true anomaly $\theta - \theta_0$. Expanding the right-hand side of this as a power series in the eccentricity, we find

$$\cos\psi = \cos(\theta - \theta_0) + e\sin^2(\theta - \theta_0) - e^2\cos(\theta - \theta_0)\sin^2(\theta - \theta_0) + \cdots$$

which yields

[6]A convenient and inexpensive source of data is *The Old Farmer's Almanac*, (Yankee Publishing Inc., Dublin, NH). A more conventional source is *The Astronomical Almanac*, (US Government Printing Office, Washington, DC).

$$\psi = (\theta - \theta_0) - e\sin(\theta - \theta_0) + \tfrac{1}{4}e^2 \sin 2(\theta - \theta_0) + \cdots .$$

This, when substituted into Kepler's equation, then gives the relation between the observed quantities, the true anomaly and the time (in years),

$$(\theta - \theta_0) - 2e\sin(\theta - \theta_0) + \tfrac{3}{4}e^2 \sin 2(\theta - \theta_0) + \cdots = 2\pi(t - t_0).$$

Setting $\theta = 0$ at autumnal equinox, $\theta = \pi/2$ at winter solstice, $\theta = \pi$ at vernal equinox, and $\theta = 3\pi/2$ at summer solstice, we obtain the four equations

$$\begin{aligned}
-\theta_0 + 2e\sin\theta_0 - \tfrac{3}{4}e^2 \sin 2\theta_0 + \cdots &= 2\pi(t_{AE} - t_0) \\
\tfrac{\pi}{2} - \theta_0 - 2e\cos\theta_0 + \tfrac{3}{4}e^2 \sin 2\theta_0 + \cdots &= 2\pi(t_{WS} - t_0) \\
\pi - \theta_0 - 2e\sin\theta_0 - \tfrac{3}{4}e^2 \sin 2\theta_0 + \cdots &= 2\pi(t_{VE} - t_0) \\
\tfrac{3\pi}{2} - \theta_0 + 2e\cos\theta_0 + \tfrac{3}{4}e^2 \sin 2\theta_0 + \cdots &= 2\pi(t_{SS} - t_0)\ .
\end{aligned}$$

These can be combined to give

$$\begin{aligned}
\frac{2e}{\pi}\sin\theta_0 &= \frac{1}{2} - (t_{VE} - t_{AE}) = \frac{1}{2} - (\text{Autumn} + \text{Winter}) \\
-\frac{2e}{\pi}\cos\theta_0 &= \frac{1}{2} - (t_{SS} - t_{WS}) = \frac{1}{2} - (\text{Winter} + \text{Spring}) \\
4t_0 &= t_{AE} + t_{WS} + t_{VE} + t_{SS} + \frac{2\theta_0}{\pi} - \frac{3}{2} \\
\frac{1}{2} + \frac{3e^2}{2\pi}\sin 2\theta_0 &= t_{WS} - t_{AE} + t_{SS} - t_{VE} = \text{Autumn} + \text{Spring}\ .
\end{aligned}$$

The first two equations give the eccentricity and angle of perihelion; the third gives the time of perihelion; and the fourth is a check on the consistency of the data with the assumption of a Keplerian orbit. In particular, using the data in Table I we find

$$\begin{aligned}
\frac{2e}{\pi}\sin\theta_0 &= 0.010385 \\
-\frac{2e}{\pi}\cos\theta_0 &= 0.002369
\end{aligned}$$

which yield $e = 0.016732$ and $\theta_0 = 102.85°$. The third equation then gives

$$4t_0 = 1604.4792 + (2 \times 102.85/180 - 1.5) \times 365.2458$$

which yields $t_0 = 368.50 = 3$ January 1995, 12h. Finally, the fourth equation gives

$$0.499942 \approx 0.499937.$$

The equations we have been using are good to order $e^2$; corrections of order $e^3 = 4.7 \times 10^{-6}$ are expected to modify the results by a few parts in the last figure quoted. The uncertainty in the input times (assumed to be $\pm 1\,\mathrm{minute} = \pm 1.9 \times 10^{-6}\,\mathrm{year}$) also affects the last figure.

The agreement with the almanac values $e = 0.01673 \pm 0.00002$ and $\theta_0 = 102.87° \pm 0.08°$, where the $\pm$ indicates the variation over the course of the year due to various perturbations, is excellent. However, our predicted $t_0$ is 18 hours early. The reason for this is interesting. We have actually been calculating the parameters of the earth-moon barycenter; in particular, our $t_0$ is the time of *barycenter* perihelion. What is listed in almanacs, however, is the time of *earth* perihelion. These differ by approximately $1.3 \sin\phi$ days, where $\phi$ is the (angular) phase of the moon near perihelion. In 1995 perihelion occurred about a third the way through the first quarter of the moon, so $\phi \approx 30°$ and the correction to our result is approximately +16 hours, as required.

## Scattering

One of the primary ways for exploring the nuclear and sub-nuclear world is via scattering experiments. A beam of particles, such as electrons, protons, or $\alpha$-particles, is directed at a thin target which contains the nucleus to be studied. The particles in the beam interact with the target nuclei and are scattered. A detector counts the number of particles per unit time scattered in various directions (Fig. 1.14(a)). This number is proportional to the (small) cross-sectional area $\Delta A$ of the detector and is inversely proportional to the square of the distance R from the target to the detector. That is, the number $\Delta n$ of counts per unit time is proportional to the **solid angle** $\Delta\Omega = \Delta A/R^2$ subtended by the detector. Further, it is proportional to the **intensity** I of the incident beam (the number of incident particles per unit area per unit time) and to the number N of target nuclei in the path of the beam. To obtain a quantity from which these details of the experimental arrangement have been removed, we divide the number of counts per unit time by the solid angle subtended by the detector, by the intensity of the incident beam, and by the number of target nuclei. The resulting quantity,

$$\frac{d\sigma}{d\Omega} = \frac{1}{NI}\frac{dn}{d\Omega},$$

is called the **differential scattering cross section**. It has units of "area": $\Delta\sigma = (d\sigma/d\Omega)\Delta\Omega$ is the area an incident particle must strike, per target nucleus, in order to scatter into the solid angle $\Delta\Omega$.

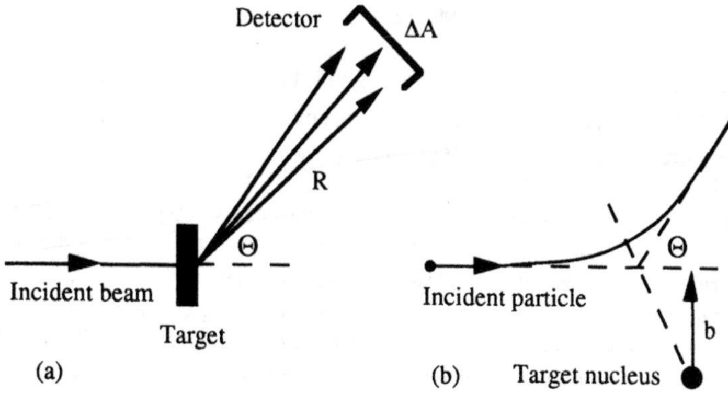


Fig. 1.14. Scattering (a) macroscopic view, (b) microscopic view

For thin targets the scattering of an incident particle is the result of a single collision between it and an individual target nucleus. We wish to relate the differential scattering cross section to this individual scattering process. The proper way to do this is to use quantum mechanics. It is nevertheless of interest to see how one approaches such a problem using classical mechanics, if only to introduce general ideas and to provide results which can then be compared with the quantum mechanical results. When the incident particle is far from the target nucleus, the force exerted on it by the target nucleus is small, and the particle moves along an incoming asymptotic straight line (Fig. 1.14(b)). If extended, this straight line would pass by the target nucleus with distance of closest approach b. This distance b is called the **impact parameter**. It is related to the (constant) angular momentum L and energy E of the incident particle by

$$L = mv_\infty b = b\sqrt{2mE}\,.$$

As a result of its interaction with the target nucleus, the incident particle is deflected from its original path, eventually emerging from the interaction region along an outgoing asymptotic straight line which makes an angle $\Theta$ $(0 \le \Theta \le \pi)$ with the incident direction. This angle $\Theta$ is called the **scattering angle**. It is a (single-valued) function $\Theta(b,E)$ of the impact parameter b and particle energy E. If we choose a spherical polar coordinate system with origin at the target and polar axis in the direction of the incident beam, the scattering angle $\Theta$ is the polar angle of the scattered particle. The azimuthal angle $\phi$ of the particle does not change (except possibly by $\pi$). The number of particles per unit time incident with impact parameter in the range b to $b + db$ and with azimuthal angle in the range $\phi$ to $\phi + d\phi$ is $dn = I\,b\,db\,d\phi$. These particles scatter into the specific polar angle range $\Theta$ to $\Theta + d\Theta$ where $\Theta$ is determined by b, and into the azimuthal angle range $\phi\,(+\pi)$ to $\phi + d\phi\,(+\pi)$. The solid angle which they subtend is thus $d\Omega = \sin\Theta\, d\Theta\, d\phi$. Substituting these results into the above definition of the differential scattering cross section then gives

$$\frac{d\sigma}{d\Omega} = \frac{b}{\sin\Theta}\left|\frac{db}{d\Theta}\right|.$$

In writing this down we have assumed that the relationship between impact parameter and scattering angle (at fixed energy) is one to one. For some interactions this may not be the case, with more than one impact parameter yielding the same scattering angle. In such situations the above should be replaced by an appropriate sum.[7]

## Coulomb scattering

The scattering of a low energy incident charged particle by a nucleus is largely the result of the electrostatic Coulomb force between it and the nucleus. We have seen that the orbit for a particle moving in a Coulomb potential $V = -k/r$ (with $k = -q_1q_2$ where $q_1$ and $q_2$ are the electric charges of the incident and target particles) is a conic section

$$\frac{p}{r} = 1 + e\cos\theta$$

where $p = L^2/mk$ is the semi-latus-rectum and $e = \sqrt{1 + 2L^2E/mk^2} = \sqrt{1 + (2bE/k)^2}$ is the eccentricity. Also, for convenience we have here taken pericenter in the direction $\theta = 0$ for an attractive force (unlike charges), or in the direction $\theta = \pi$ for a repulsive force (like charges). See Fig. 1.15.

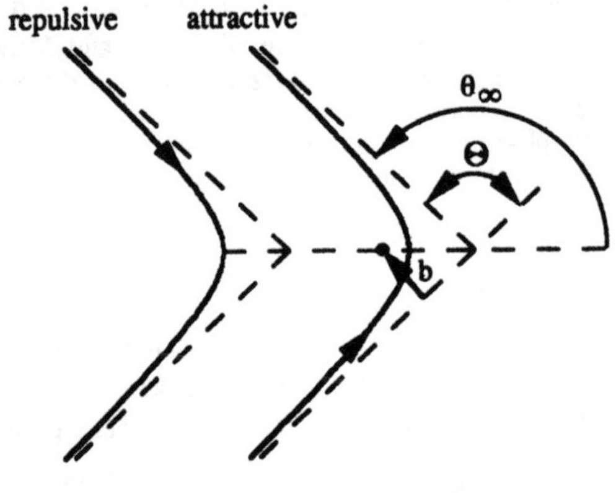


Fig. 1.15. Coulomb scattering

[7]For some of the things that can happen then see: Roger G. Newton, *Scattering Theory of Waves and Particles*, (McGraw-Hill Book Company, New York, NY, 1966), pp. 129-134; Herbert Goldstein, *Classical Mechanics*, (Addison-Wesley Publishing Company, Reading, MA, 1980), 2nd. ed., pp. 110-113.

For E positive, e is greater than one, and the orbit is a hyperbola. Far from the nucleus ($r \to \infty$) the incident particle travels along incoming or outgoing asymptotic straight lines, with directions $\mp\theta_\infty$ where

$$\theta_\infty = \cos^{-1}(-1/e) \qquad (\pi/2 \le \theta_\infty \le \pi).$$

Note that these directions are the same (apart from sign) for both attractive and repulsive Coulomb scattering. The scattering angle is given by

$$\Theta = 2\theta_\infty - \pi.$$

This, combined with the previous equations, yields the relation between the impact parameter and the scattering angle

$$b = \frac{|k|}{2E}\cot\frac{\Theta}{2}.$$

The scattering angle $\Theta$ is a monotonely decreasing function of the impact parameter b, decreasing from $\Theta = \pi$ at $b = 0$ to $\Theta = 0$ as $b \to \infty$. The differential scattering cross section follows from previous considerations and is given by

$$\frac{d\sigma}{d\Omega} = \left(\frac{k}{4E}\right)^2 \csc^4\left(\frac{\Theta}{2}\right).$$

This is the famous **Rutherford scattering cross section**, first derived and used by Ernest Rutherford to interpret the experiments of Geiger and Marsden on $\alpha$-particle scattering from thin metal foils. It led him to the discovery of the nuclear atom. As has already been pointed out, this is really a quantum mechanical problem. Fortunately for the development of atomic physics, however, in this instance classical mechanics, more by accident than anything, leads to the same result as quantum mechanics.[8]

## Exercises

1. A particle of mass m moves in one dimension x in a potential well

$$V = V_0 \tan^2(\pi x/2a)$$

where $V_0$ and a are constants. Find, for given total energy E, the position x as a function of time and the period $\tau$ of the motion. In particular, examine and interpret the low energy ($E \ll V_0$) and high energy ($E \gg V_0$) limits of your expressions.

---

[8]See, for example: Kurt Gottfried, *Quantum Mechanics*: Volume I, (W. A. Benjamin, New York, NY, 1966), pp. 148-153; Gordon Baym, *Lectures on Quantum Mechanics*, (W. A. Benjamin, Reading, MA, 1969, 1973), pp. 213-224; J. J. Sakurai, *Modern Quantum Mechanics*, (Addison-Wesley, Redwood City, CA, 1985), ed. San Fu Tuan, pp. 434-444.

2. For each of the following central potentials $V(r)$ sketch the effective potential

$$V_{eff}(r) = \frac{L^2}{2mr^2} + V(r),$$

and use your sketch to classify and draw qualitative pictures of the possible orbits.

(a) $V(r) = \frac{1}{2}kr^2$ **3D isotropic harmonic oscillator**

(b) $V(r) = -V_1$ for $r < a$ **square well**
$V(r) = 0$ for $r > a$

(c) $V(r) = -\frac{k}{r^2}$

(d) $V(r) = -\frac{k}{r^4}$

(e) $V(r) = -k\frac{e^{-\alpha r}}{r}$ **Yukawa potential**

Note that the qualitative shape of $V_{eff}(r)$ versus r may depend on L and on the various parameters; consider all cases (but assume that the given parameters are positive).

3. The first U.S. satellite to go into orbit, Explorer I, which was launched on January 31, 1958, had a perigee of 360 km and an apogee of 2549 km above the earth's surface. Find:
(a) the semi-major axis,
(b) the eccentricity,
(c) the period,
of Explorer I 's orbit. The earth's equatorial radius is 6378 km and the acceleration due to gravity at the earth's surface is $g = 9.81 m/s^2$.

4. Mars travels on an approximately elliptical orbit around the Sun. Its minimum distance from the Sun is about 1.38 AU and its maximum distance is about 1.67 AU (1 **AU** = mean distance from Earth to Sun). Find:
(a) the semi-major axis,
(b) the eccentricity,
(c) the period,
of Mars' orbit.

5. The most economical method of traveling from one planet to another, the **Hohmann transfer**, consists of moving along a (Sun-controlled) elliptical path which is tangent to the (approximately) circular orbits of the two planets. Consider a Hohmann transfer from Earth (orbit radius 1.00 AU) to Venus (orbit radius 0.72 AU). Find, in units of AU and year:
(a) the semi-major axis of the transfer orbit,
(b) the time required to go from Earth to Venus,
(c) the velocity "kick" needed to place a spacecraft in Earth orbit into the transfer orbit.
In this problem ignore the effects of the gravitational fields of Earth and Venus on the spacecraft.

6. Halley's comet travels around the Sun on an approximately elliptical orbit of eccentricity $e = 0.967$ and period 76 years. Find:
(a) the semi-major axis of the orbit (Ans. 17.9 AU),
(b) the distance of closest approach of Halley's comet to the Sun (Ans. 0.59 AU),
(c) the time per orbit that Halley's comet spends within 1 AU of the Sun (Ans. 78 days).

7. Define a "season" as a time interval over which the true anomaly increases by $\pi/2$. Find the duration of the shortest season for earth. Take the eccentricity of earth's orbit to be 0.0167.

8. A satellite of mass m moves in a circular orbit of radius $a_0$ around the earth.
(a) A rocket on the satellite fires a burst radially, and as a result the satellite acquires, essentially instantaneously, a radial velocity u in addition to its angular velocity. Find the semi-major axis, the eccentricity, and the orientation of the elliptical orbit into which the satellite is thrown.
(b) Repeat (a), if instead the rocket fires a burst tangentially.
(c) In both cases find the velocity kick required to throw the satellite into a parabolic orbit.

9. Show that the following ancient picture of planetary motion (in heliocentric terms) is in accord with Kepler's picture, if the eccentricity e is small and terms of order $e^2$ and higher are neglected:
(a) the earth moves around the sun in a circular orbit of radius a; however, the sun is not at the center of this circle, but is displaced from the center by a distance ea;
(b) the earth does not move uniformly around the circle; however, a radius vector from a point which is on a line from the sun to the center, the same distance from and on the opposite side of the center as the sun, to the earth does rotate uniformly.

10. (a) Show that

$$\frac{2r_0}{r} = 1 + \cos\theta$$

(the standard form for a conic section, on setting the eccentricity $e = 1$ and the semi-latus-rectum $p = 2r_0$) is the equation of a parabola, by translating it into cartesian coordinates with the origin at the focus and the x-axis through pericenter.
(b) A comet travels around the Sun on a parabolic orbit. Show that the distance r of the comet from the Sun is related to the time t from perihelion by

$$\frac{\sqrt{2}}{3}(r + 2r_0)\sqrt{r - r_0} = 2\pi t$$

where distances are measured in AU and time is measured in years.
(c) If one approximates the orbit of Halley's comet near the Sun by a parabola with $r_0 = 0.59\,\mathrm{AU}$, what does this give for the time Halley's comet spends within 1 AU of the Sun?
(d) What is the maximum time a comet on a parabolic orbit may spend within 1 AU of the Sun?

11 A particle of mass m moves in a central force field $\mathbf{F} = -(k/r^2)\hat{\mathbf{r}}$.
(a) By integrating Newton's second law $d\mathbf{p}/dt = \mathbf{F}$, show that the momentum of the particle is given by $\mathbf{p} = \mathbf{p}_0 + (mk/L)\hat{\boldsymbol{\theta}}$, where $\mathbf{p}_0$ is a constant vector and L is the magnitude of the angular momentum.
(b) Hence show that the orbit in momentum space (the so-called **hodograph**) is a circle. Where is the center and what is the radius of the circle?
(c) Show that the magnitude of $\mathbf{p}_0$ is $(mk/L)e$, where e is the eccentricity. Sketch the orbit in momentum space for the various cases, $e = 0$, $0 < e < 1$, $e = 1$, $e > 1$, indicating for the last two cases which part of the circle is relevant.
(See: Arnold Sommerfeld, *Mechanics*, (Academic Press, New York, NY, 1952), trans. Martin O. Stern, p. 33, 40, 242; Harold Abelson, Andrea diSessa, and Lee Rudolph, "Velocity space and the geometry of planetary orbits," Am. J. Phys. **43**, 579-589 (1975).)

12. Consider the motion of a particle in a central force field with potential $V = -k/r$. Since the force is central, the angular momentum $\mathbf{L} = \mathbf{r} \times \mathbf{p}$ is constant and the orbit lies in a plane passing through the force center and perpendicular to **L**.
(a) Show that for the particular potential $V = -k/r$ there exists an additional vector quantity which is constant, the **Laplace-Runge-Lenz vector**

$$\mathbf{K} = \mathbf{p} \times \mathbf{L} - mk\hat{\mathbf{r}}.$$

Further show that $\mathbf{K} \cdot \mathbf{L} = 0$, so that **K** and **L** are perpendicular and thus **K** lies in the orbital plane. (Hint: if you've done exercise 1.11, you need only show that $\mathbf{K} = \mathbf{p}_0 \times \mathbf{L}$).
(b) By taking the dot product of **K** with $\hat{\mathbf{r}}$ obtain the equation of the orbit

$$\frac{a(1-e^2)}{r} = 1 + e\cos\theta.$$

Hence find a and e in terms of K and L, and also find the direction that **K** points in the orbital plane.
(c) Express the energy $E = \dfrac{p^2}{2m} - \dfrac{k}{r}$ in terms of K and L.

13. Consider the motion of a particle of mass m in a central force field with potential

$$V = -\frac{k}{r} + \frac{h}{r^2}.$$

(a) Show that the equation for the orbit can be put in the form

$$\frac{a(1-e^2)}{r} = 1 + e\cos\alpha\theta,$$

and find a, e, and $\alpha$ in terms of the energy E and angular momentum L of the particle, and the parameters k and h of the potential.
(b) Show that this represents a precessing ellipse, and derive an expression for the average rate of precession in terms of the dimensionless quantity $\eta = h/ka$.
(c) The perihelion of Mercury precesses at the rate of 40″ of arc per century, after all known planetary perturbations are taken into account. What value of $\eta$ would

lead to this result? The eccentricity of Mercury's orbit is 0.206 and its period is 0.24 years.

14. A particle of mass m moves in a 3D isotropic harmonic oscillator potential well

$$V = \tfrac{1}{2} m\omega^2 r^2$$

where $\omega$, the angular frequency, is a constant.

(a) Show that the equation for the orbit has the form

$$\frac{L^2}{mE}\frac{1}{r^2} = 1 + \sqrt{1 - \frac{\omega^2 L^2}{E^2}} \cos 2(\theta - \theta_0)$$

where E is the energy and L is the angular momentum.

(b) Show that this represents an ellipse with geometric center at the force center, and express the energy and angular momentum in terms of the semi-major axis a and eccentricity e of the ellipse. (Ans. $E = m\omega^2(a^2 + b^2)$ and $L = m\omega ab$ where $b = a\sqrt{1 - e^2}$ is the semi-minor axis)

(c) Show that the period is $\tau = 2\pi/\omega$ independent of the energy and angular momentum, and that the radius is given as a function of time by

$$r^2 = \frac{E}{m\omega^2}\left[1 - \sqrt{1 - \frac{\omega^2 L^2}{E^2}} \cos 2\omega(t - t_0)\right].$$

15. A small meteor approaches the earth with impact parameter b and velocity $v_\infty$ at infinity. Show that the meteor will strike the earth if

$$b < a\sqrt{1 + (v_0/v_\infty)^2}$$

where a is the radius and $v_0$ is the "escape velocity" for the earth.

16.

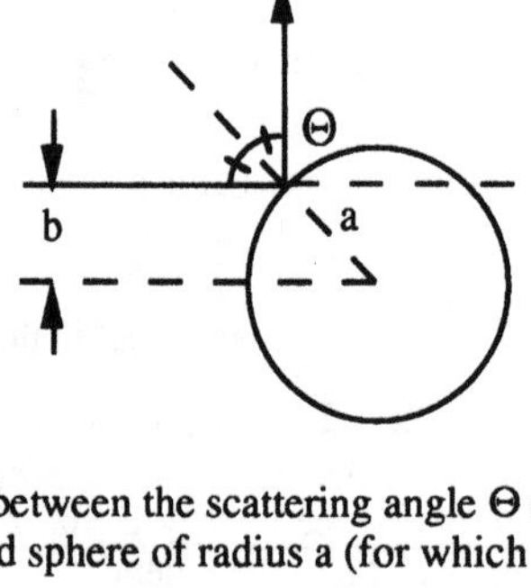


(a) Find the relation between the scattering angle Θ and the impact parameter b for scattering from a hard sphere of radius a (for which "angle of incidence = angle of reflection").

(b) Use your result to obtain the differential scattering cross section $d\sigma/d\Omega$. Integrate to find the **total scattering cross section** $\sigma = \int (d\sigma/d\Omega)d\Omega$, where the integration extends over the whole solid angle.

17. (a) Show that a particle of energy E is refracted in going from a region in which the potential is zero to a region in which the potential is $-V_1$, the angle of incidence $\theta_0$ and the angle of refraction $\theta_1$ being related by **Snell's law**

$$\frac{\sin\theta_0}{\sin\theta_1} = n$$

where angles are measured from the normal and $n = \sqrt{1 + V_1/E}$ is the **index of refraction.**

(b) Use Snell's law to show that a particle incident at impact parameter b on an attractive square well potential

$$V(x) = -V_1 \quad \text{for } r < a$$
$$V(x) = 0 \quad \text{for } r > a$$

is scattered through an angle $\Theta$ given by

$$\frac{b^2}{a^2} = \frac{n^2 \sin^2 \Theta/2}{n^2 + 1 - 2n\cos\Theta/2}.$$

In particular, show that for small impact parameters ($b << a$) the scattered particles are brought to a focus a distance $f \approx \left(\frac{n}{n-1}\right)\left(\frac{a}{2}\right)$ from the force center.

(c) Find the differential scattering cross section $d\sigma/d\Omega$.

18. (a) Show that

$$\frac{r_0}{r} = \cos\alpha\theta$$

is the equation of the orbit for a particle moving in a repulsive potential $V(r) = k/r^2$, determining $\alpha$ and $r_0$ in terms of the energy and angular momentum.

(Ans. $\alpha = \sqrt{1 + \frac{2mk}{L^2}}$, $r_0 = \frac{\alpha L}{\sqrt{2mE}}$)

(b) Show that the impact parameter b and scattering angle $\Theta$ are related by

$$b^2 = \frac{k}{E}\frac{(\pi - \Theta)^2}{\Theta(2\pi - \Theta)}.$$

(c) Show that the differential scattering cross section is given by

$$\frac{d\sigma}{d\Omega} = \frac{\pi^2 k}{E\sin\Theta}\frac{\pi - \Theta}{\Theta^2(2\pi - \Theta)^2}.$$

# CHAPTER II

# THE PRINCIPLE OF VIRTUAL WORK AND D'ALEMBERT'S PRINCIPLE

In the previous chapter we saw how Newton's laws could be used directly to solve some simple mechanical problems involving point particles. We now turn to more general mechanical systems. We shall see that for most mechanical systems Newton's laws are *incomplete* and must be supplemented by additional conditions. These are contained in the principle of virtual work which is the subject of the present chapter.[1]

## Constraints

We begin by writing down Newton's second law as applied to a system of N particles,

$$m_i \ddot{\mathbf{r}}_i = \mathbf{F}_i \qquad i = 1,2,\cdots,N.$$

At first it might appear that all we have to do is to integrate this coupled set of 3N equations to obtain the 3N coordinates $\mathbf{r}_i$ as functions of time. We soon discover, however, apart from the fact that the integration is unfeasible in most situations, that this set of equations is incomplete. There is more to mechanics than Newton's second law. In particular, the coordinates might be related or restricted by **constraints**. For example:

(a) The particles might be required to move on certain surfaces or curves, as for a block sliding on an inclined plane, or for a plane pendulum (Fig. 2.01).

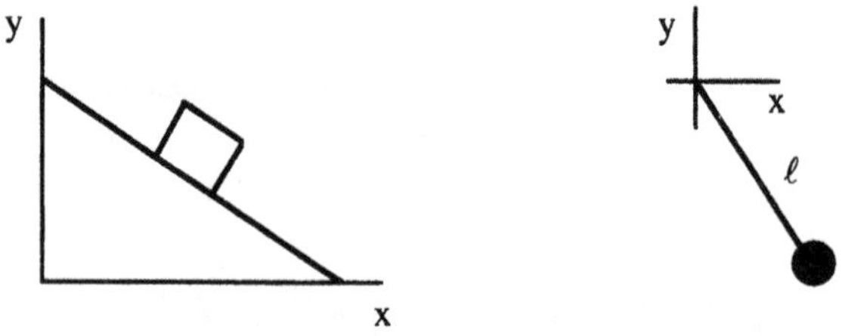


The block moves on the surface $y = ax + b$

The bob moves on the curve $x^2 + y^2 = \ell^2$, $z = 0$

Fig. 2.01. Typical constraints

[1]For parallel reading see: Robert A. Becker, *Introduction to Theoretical Mechanics*, (McGraw-Hill Book Company, New York, NY, 1954), pp. 97-107; Cornelius Lanczos, *The Variational Principles of Mechanics*, (University of Toronto Press, Toronto, Ont., 1970; republished by Dover Publications, New York, NY, 1986), 4th ed., pp. 74-110; Arnold Sommerfeld, *Mechanics*, (Academic Press, New York, NY, 1952), trans. Martin O. Stern, pp. 48-66.

(b) The particles might be connected by "light rigid rods" so that the distances between them remain constant,

$$|\mathbf{r}_i - \mathbf{r}_j| = a_{ij},$$

as for the particles which make up a rigid body.

Constraints such as these which can be expressed as a set of equations of the form

$$G_l(\mathbf{r}_1, \mathbf{r}_2, \cdots, \mathbf{r}_N; t) = 0 \qquad l = 1, 2, \cdots, 3N - f$$

are called **holonomic** constraints. The integer "f" is the number of **degrees of freedom** of the system. Other (non-holonomic) types of constraint, such as "particles confined to the interior of a box" or "wheel rolling over a surface," are difficult to handle in a general way and are not considered here.

Constraints have two effects:

1. The 3N coordinates $\mathbf{r}_i = (x_i, y_i, z_i)$ are not all independent. For a system with f degrees of freedom there are only f independent coordinates.

2. There are **forces of constraint** $\mathbf{F}_i^{(\text{constraint})}$ which the constraining surfaces, curves, rods, etc. exert on the particles so that they move in conformity with the constraints. These are initially *unknown* and must be found as part of the solution to the problem. If we call all the other forces **applied forces** and denote them by $\mathbf{F}_i^{(\text{applied})}$, the 3N equations arising from Newton's second law take the form

$$m_i \ddot{\mathbf{r}}_i = \mathbf{F}_i^{(\text{applied})} + \mathbf{F}_i^{(\text{constraint})} \qquad i = 1, 2, \cdots, N.$$

Together with the equations of constraint, these give a total of $6N - f$ equations for the 6N unknowns $\mathbf{r}_i$ and $\mathbf{F}_i^{(\text{constraint})}$. Thus at the moment we do not have sufficient information to solve the dynamical problem and must impose further conditions. In order to discover what these might be, let us look at some elementary examples where this situation occurs and see what we do in those cases.

## Principle of virtual work

First consider a block sliding on a frictionless incline near the surface of the earth (Fig. 2.02).

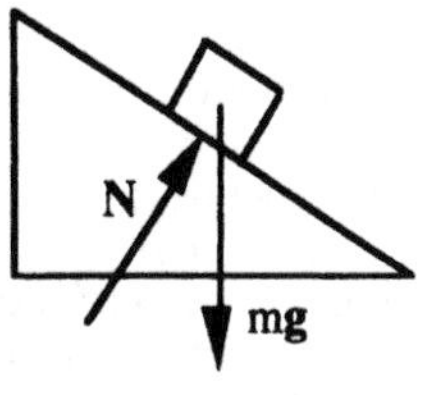


Fig. 2.02. Block on an incline

The block is subject to two forces: the force $\mathbf{mg}$ gravity exerts on the block, an applied force, and the force $\mathbf{N}$ the incline exerts on the block, a force of constraint. If we think of this as a problem in two dimensions, there are four unknowns $x$, $y$, $N_x$, $N_y$. To find these, we have two equations from Newton's second law and one equation of constraint. The necessary fourth equation is the statement that $\mathbf{N}$ is perpendicular to the incline. We now wish to say this in a way which can be applied to a wide variety of situations. For this purpose we observe that another way to obtain the same result is to say that the force of constraint, being perpendicular to the displacement, does no work. We shall see that this idea, suitably extended, leads to the general additional condition we must impose on a mechanical system so as to make a well-set problem.

Next consider two particles connected by a light rigid rod and possibly subjected to external forces (Fig. 2.03).

Fig. 2.03. Two interacting particles

We want to find the coordinates $\mathbf{r}_1$, $\mathbf{r}_2$ of the particles and the constraint forces $\mathbf{F}_1$, $\mathbf{F}_2$ the rod exerts on them, twelve unknowns in all. We have six equations from Newton's second law and one constraint equation. The remaining necessary equations are

$$\mathbf{F}_1 = -\mathbf{F}_2 \qquad \text{(3 equations)},$$

the force the rod exerts on particle 1 is equal and opposite the force it exerts on particle 2, and

"the forces are directed along a line joining the two particles" (2 equations).

How are we to summarize conveniently these requirements? We observe that for any displacement of the system, while the forces of constraint $\mathbf{F}_1$ and $\mathbf{F}_2$ may do work individually, the *net* work

$$\delta W = \mathbf{F}_1 \cdot \delta \mathbf{r}_1 + \mathbf{F}_2 \cdot \delta \mathbf{r}_2$$

done by all the forces of constraint is again zero. To see this, note that the displacements are of two types: *translations*, for which $\delta \mathbf{r}_1 = \delta \mathbf{r}_2$, and $\delta W = 0$ because $\mathbf{F}_1 = -\mathbf{F}_2$ and the work done by $\mathbf{F}_1$ is equal and opposite the work done by $\mathbf{F}_2$; and *rotations*, for which the displacements are perpendicular to the line joining the two particles, and the work done by $\mathbf{F}_1$ and $\mathbf{F}_2$ are each zero because the forces lie along the line joining the two particles. Further, by reversing the argument we can obtain the preceding five conditions on the forces of constraint from the statement about work; they are equivalent.

As we continue to examine a wide variety of situations, we may be tempted to summarize our observations by saying "the net work done by forces of constraint is zero," but this is not quite true. Forces of constraint *can* do work if the constraint is time-dependent, if the incline in the first example is moving or the length of the rod in the second example is changing. Consider Fig. 2.04,

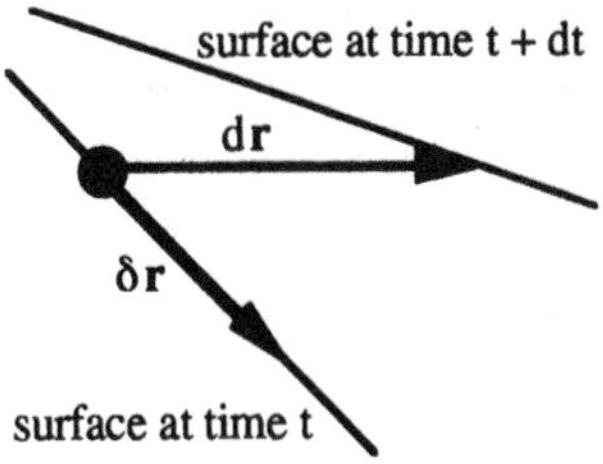


Fig. 2.04. Real and virtual displacements for a time-varying constraint

which shows a particle constrained to a surface. If the surface moves in the time interval t to t + dt, the real displacement **dr** of the particle has a component normal to the surface, in the direction of the force of constraint, so the force of constraint in this situation can do work. Thus in order to apply our work idea we must modify our prescription as follows: *freeze* the system at some instant of time t ; then *imagine* the particles displaced amounts $\delta \mathbf{r}_i$ consistent with the conditions of constraint. This is called a **virtual displacement.** We use $\delta \mathbf{r}_i$ rather than $d\mathbf{r}_i$ to distinguish virtual displacements from real displacements. We then apply our work idea not to real displacements but to virtual displacements, stating the result as follows:

**The principle of virtual work**

The total work done by the forces of constraint in a virtual displacement is zero,

$$\delta W^{(\text{constraint})} = \sum_{i=1}^{N} \mathbf{F}_i^{(\text{constraint})} \cdot \delta \mathbf{r}_i = 0.$$

We have not given a "proof" of the principle of virtual work, but rather an indication of some types of situation in which the principle holds. Readers will have to judge from physical considerations whether and in what sense the principle holds for the particular physical system they wish to consider. The principle is in a sense a statement about what forces we can consider "forces of constraint," and it summarizes their properties. Forces which do not satisfy the principle must be considered "applied forces."

As we now show, the principle of virtual work provides the additional f equations needed, besides the 3N from Newton's second law and the 3N − f equations of constraint, to complete the specification of the dynamical problem. First suppose that there are no constraints. Then all the $\delta \mathbf{r}_i$'s are independent, and the only way $\delta W^{(\text{constraint})}$ can be zero for all $\delta \mathbf{r}_i$ is if $\mathbf{F}_i^{(\text{constraint})} = 0$; these 3N equations say, correctly, that in this case there are no forces of constraint. Now suppose that there is one constraint. The coordinates are then connected by one equation of the form

$$G(\mathbf{r}_1, \mathbf{r}_2, \cdots, \mathbf{r}_N; t) = 0,$$

and the number of degrees of freedom is $3N-1$. That is, $3N-1$ of the $\mathbf{r}_i = (x_i, y_i, z_i)$ are independent, and 1 is dependent. If we express this one dependent variable in terms of the independent variables $x_j^{(ind)}$, the principle of virtual work gives

$$\sum_{j=1}^{3N-1}\left(\sum_{i=1}^{N}\mathbf{F}_i^{(constraint)}\cdot\frac{\partial\mathbf{r}_i}{\partial x_j^{(ind)}}\right)\delta x_j^{(ind)} = 0.$$

The coefficient of each of the $\delta x_j^{(ind)}$ must vanish, giving $3N-1$ restrictions on the $\mathbf{F}_i^{(constraint)}$'s, as required. It is clear that this argument can be easily generalized to the case where there are $3N-f$ equations of constraint and f independent variables. Each time we add a constraint equation we reduce the number of degrees of freedom, the number of independent variables, by one and hence reduce the number of conditions on the $\mathbf{F}_i^{(constraint)}$ by one; the number of constraint equations plus conditions on the $\mathbf{F}_i^{(constraint)}$ remains fixed at 3N. To summarize, the principle of virtual work provides the additional equations needed to make a well-set mechanical problem.

## D'Alembert's principle and generalized coordinates

Quite often we are not interested in the forces of constraint themselves. We can then use Newton's second law to eliminate them from the remaining equations, setting

$$\mathbf{F}_i^{(constraint)} = m_i\ddot{\mathbf{r}}_i - \mathbf{F}_i^{(applied)}$$

in the principle of virtual work. We are left with

$$\sum_{i=1}^{N}\left(\mathbf{F}_i^{(applied)} - m_i\ddot{\mathbf{r}}_i\right)\cdot\delta\mathbf{r}_i = 0.$$

This is **d'Alembert's principle**. It says that the work done by the applied forces, plus the work done by the so-called **inertial forces** $-m_i\ddot{\mathbf{r}}_i$, in a virtual displacement is zero. In spite of its simple appearance, d'Alembert's principle is extremely important. Together with the equations of constraint, it *contains* Newton's second law as well as the necessary conditions on the forces of constraint. It forms the basis for all our further developments in mechanics.

Rather than using a set of 3N *non-independent* variables $\mathbf{r}_i$ which are connected by the $3N-f$ equations of constraint, it is more convenient to use a set of f ($\leq 3N$) *independent* variables $q_a$ ($a = 1,2,\cdots,f$), the **generalized coordinates**, to describe the configuration of the system. We have great freedom in the choice of these coordinates. Essentially any set of f independent variables from which the $\mathbf{r}_i$ can be obtained by equations of the form

$$\mathbf{r}_i = \mathbf{r}_i(q_1, q_2, \cdots, q_f; t) \qquad i = 1, 2, \cdots, N$$

will serve. See, for example, Fig. 2.05.

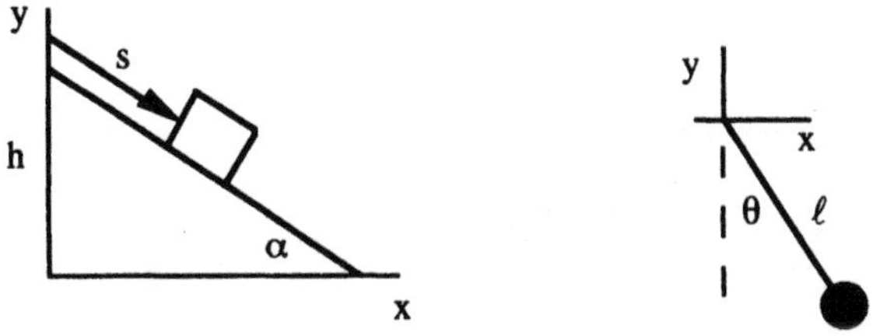


(a) For the block on an inclined plane the horizontal coordinate x *or* the vertical coordinate y *or* the distance s down the plane would all serve as the generalized coordinate. In terms of the latter variable the cartesian coordinates are

$x = s\cos\alpha \qquad y = h - s\sin\alpha$

(b) For the plane pendulum the horizontal coordinate x *or* the vertical coordinate y *or* the angle θ would all serve as the generalized coordinate. In terms of the latter variable the cartesian coordinates are

$x = \ell\sin\theta \qquad y = -\ell\cos\theta$

Fig. 2.05. Typical generalized coordinates

Once we have introduced generalized coordinates for a system, the dynamics is completely contained in d'Alembert's principle. Let us see how to use it for some simple problems in mechanics.

## Lever

A (horizontal) lever is in static equilibrium under the application of (vertical) forces $F_1$ a distance $\ell_1$ from the fulcrum, and $F_2$ a distance $\ell_2$ from the fulcrum, as shown in Fig. 2.06.

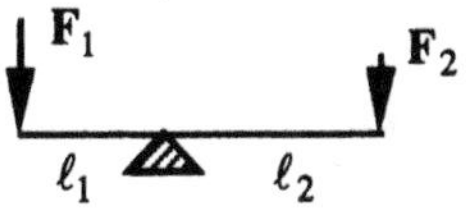


Fig. 2.06. Lever

What is the condition on these quantities for equilibrium to obtain? Although the answer is well-known to any child who has experimented with a teeter-totter, obtaining it via d'Alembert's principle is instructive. To do this, imagine the lever to undergo a virtual displacement, a (say) clockwise rotation about its fulcrum through an infinitesimal angle

$\delta\theta$. End 1 moves up a distance $\ell_1\delta\theta$ and the applied force $F_1$ does work $-F_1\ell_1\delta\theta$; end 2 moves down a distance $\ell_2\delta\theta$ and the applied force $F_2$ does work $+F_2\ell_2\delta\theta$. In this problem of static equilibrium there are no inertial forces, so d'Alembert's principle yields

$$-F_1\ell_1\delta\theta + F_2\ell_2\delta\theta = 0,$$

which gives the well-known condition

$$F_1\ell_1 = F_2\ell_2.$$

## Inclined plane

A block of mass m slides on an inclined plane under the influence of gravity. We take as generalized coordinate the displacement s down the plane (Fig. 2.07).

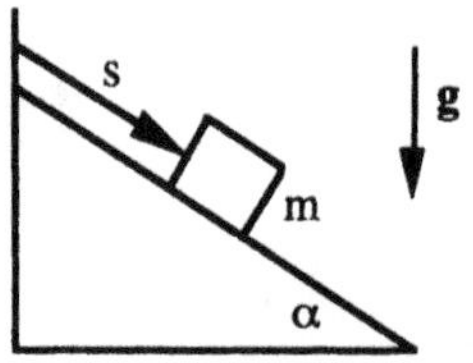


Fig. 2.07. Inclined plane

What is the equation of motion? To apply d'Alembert's principle, imagine that the block undergoes a virtual displacement $\delta s$ down the plane. The applied force, gravity, does work $mg\sin\alpha\,\delta s$. The acceleration of the block down the plane is $\ddot{s}$, so the inertial force on it is $m\ddot{s}$ up the plane, and the work done by the inertial force is $-m\ddot{s}\delta s$. D'Alembert's principle then says

$$mg\sin\alpha\,\delta s - m\ddot{s}\delta s = 0,$$

which yields the well-known result

$$\ddot{s} = g\sin\alpha.$$

## Plane pendulum

A bob of mass m is suspended from the ceiling by a string[2] of length $\ell$ and can swing back and forth in a vertical plane under the influence of gravity g. The system has

[2]Although the word "string" is used here and in other similar situations throughout the text, the phrase "light rigid rod" is sometimes meant.

one degree of freedom, and we can take as generalized coordinate the angular displacement $\theta$ from vertical (Fig. 2.08).

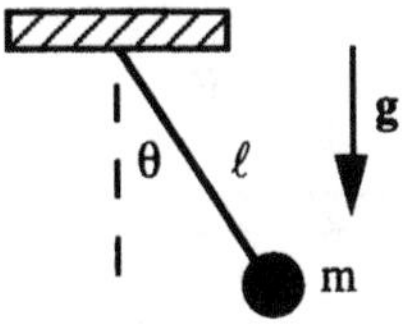


Fig. 2.08. Plane pendulum

What is the equation of motion? To apply d'Alembert's principle, imagine a virtual displacement in which $\theta$ increases by a small amount $\delta\theta$. The bob rises a distance $\ell\delta\theta\sin\theta$, and the only applied force, gravity, does work $-(mg)(\ell\delta\theta\sin\theta)$. The acceleration of the bob in the direction of the virtual displacement is $\ell\ddot{\theta}$, so the work done by the inertial force is $(-m\ell\ddot{\theta})(\ell\delta\theta)$. D'Alembert's principle then gives

$$(-mg)(\ell\delta\theta\sin\theta) - (m\ell\ddot{\theta})(\ell\delta\theta) = 0,$$

which simplifies to

$$\ddot{\theta} = -(g/\ell)\sin\theta.$$

This is the required equation of motion.

Now suppose that the length of the supporting string is time-dependent; perhaps it is expanding and contracting with changes in temperature. A virtual displacement at time t is the same as before, a distance $\ell(t)\delta\theta$ in the $\theta$-direction, so the work done by gravity is the same. But now the acceleration of the bob has a component $\ell\ddot{\theta} + 2\dot{\ell}\dot{\theta}$ in the $\theta$-direction, so the work done by the inertial force is $-m(\ell\ddot{\theta} + 2\dot{\ell}\dot{\theta})\ell\delta\theta$. D'Alembert's principle gives

$$-(mg)(\ell\delta\theta\sin\theta) - m(\ell\ddot{\theta} + 2\dot{\ell}\dot{\theta})\ell\delta\theta = 0,$$

which yields

$$\frac{d}{dt}(m\ell^2\dot{\theta}) = -mg\ell\sin\theta.$$

The quantity $m\ell^2\dot{\theta}$ is the angular momentum of the bob about the point of support. If $g = 0$, in which case the plane pendulum becomes a plane rotator, it remains constant even if $\ell$ changes with time (in contrast to, say, the kinetic energy).

Another way in which the length of the pendulum could change with time would be for the string to pass through a small hole in the ceiling and be acted on by a force F (Fig. 2.09).

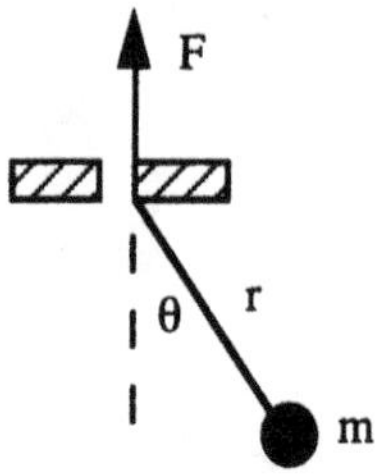


Fig. 2.09. Plane pendulum with time-varying length

The system now has *two* degrees of freedom, and we can take as generalized coordinates the angle $\theta$ and the length r of the pendulum (replacing $\ell$). There are two independent virtual displacements:

(a) Vary $\theta$, keeping r fixed. This is the same as we had in the previous paragraph, and d'Alembert's principle yields in the new notation

$$\frac{d}{dt}(mr^2\dot{\theta}) = -mgr\sin\theta .$$

(b) Vary r, keeping $\theta$ fixed. Imagine increasing r an amount $\delta r$. The applied force gravity does work $mg\delta r\cos\theta$. The applied force F does work $-F\delta r$. The acceleration of the bob has a component $\ddot{r} - r\dot{\theta}^2$ in the r-direction, so the work done by the inertial force is $-m(\ddot{r} - r\dot{\theta}^2)\delta r$. D'Alembert's principle gives

$$mg\delta r\cos\theta - F\delta r - m(\ddot{r} - r\dot{\theta}^2)\delta r = 0,$$

which yields

$$m(\ddot{r} - r\dot{\theta}^2) = -F + mg\cos\theta .$$

For $g = 0$ these are simply the general equations for motion under a central force F.

## Exercises

1.

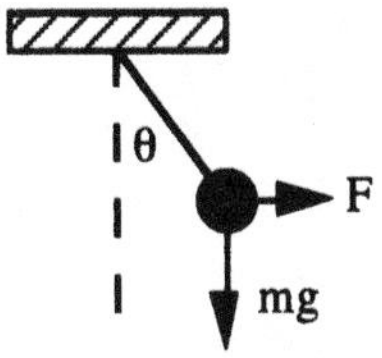


Use d'Alembert's principle to find the condition of static equilibrium.

2.

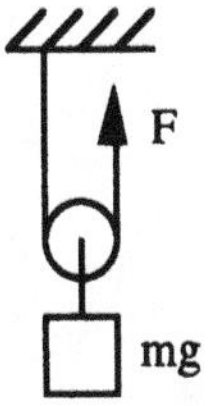


Use d'Alembert's principle to find the condition of static equilibrium.

3.

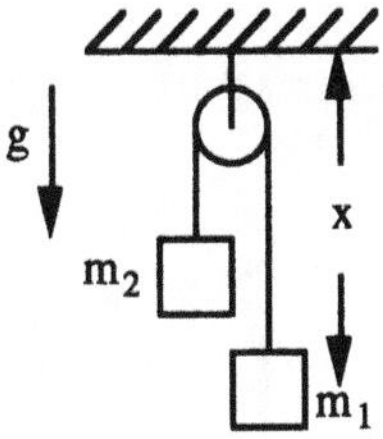


Use d'Alembert's principle to find the acceleration of $m_1$.

4.

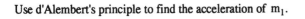

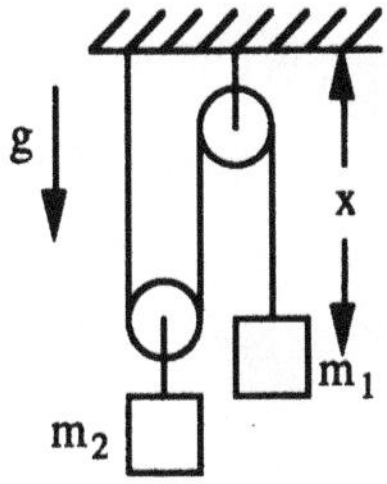


Use d'Alembert's principle to find the acceleration of $m_1$.

5.

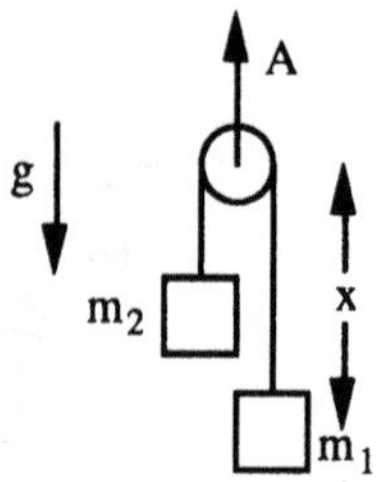


Use d'Alembert's principle to find the acceleration of $m_1$. Note that in this case the pulley has an upward acceleration A. "Acceleration" means "acceleration relative to the earth."

6.

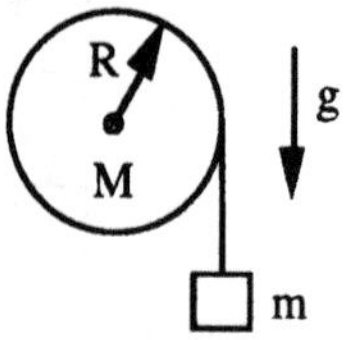


A mass m is attached to a light cord which wraps around a frictionless pulley of mass M, radius R, and moment of inertia $I = \int r^2 dM$. Gravity g acts vertically downwards. Use d'Alembert's principle to find the acceleration of m.

7.

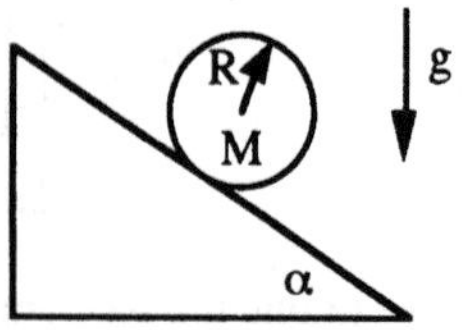


A cylinder of mass M, radius R, and moment of inertia $I = \int r^2 dM$ rolls without slipping down an inclined plane. Use d'Alembert's principle to find the acceleration of the cylinder.

8.

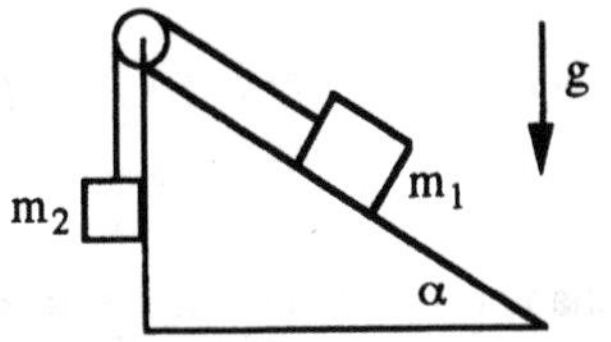


Use d'Alembert's principle to find the acceleration of $m_1$ down the (stationary) plane.

9.

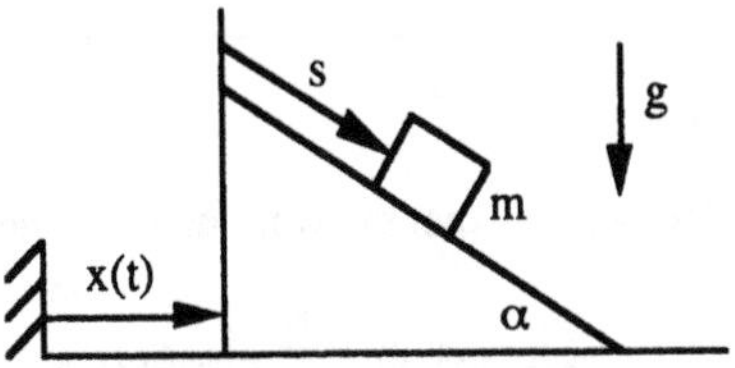


A block of mass m slides on a frictionless inclined plane, which is driven so that it moves horizontally, the displacement of the plane at time t being some known function x(t). Use d'Alembert's principle to find the equation of motion of the block, taking as generalized coordinate the displacement s of the block down the plane. Note that the acceleration of the block is *not* "down the plane."

10.

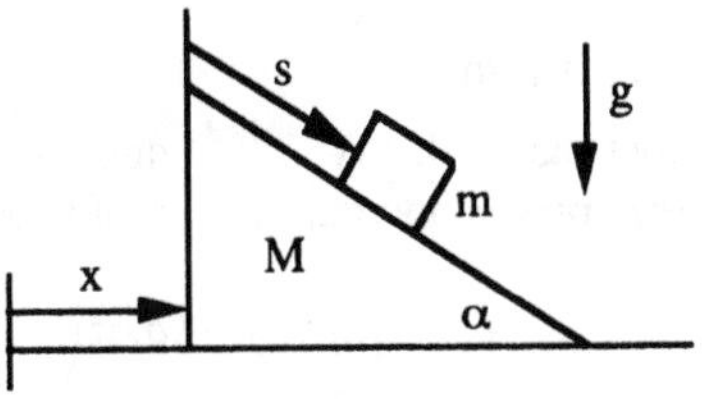


A block of mass m slides on a frictionless inclined plane of mass M which in turn is free to slide on a frictionless horizontal surface. Use d'Alembert's principle to find the equations of motion of the block and the plane, taking as generalized coordinates the displacement s of the block down the plane and the horizontal displacement x of the plane.

# CHAPTER III

# LAGRANGE'S EQUATIONS

In the preceding chapter we obtained d'Alembert's principle and showed how it could be used to find the equations of motion. This approach is convenient in that the usually uninteresting forces of constraint do not need to be considered. In this chapter we extend these ideas and show how, starting from d'Alembert's principle, we can write the equations of motion for a wide range of mechanical systems, described by arbitrary sets of generalized coordinates, in an elegant and compact way known as Lagrange's equations.

## Lagrange's equations

D'Alembert's principle says that the work done by the applied forces

$$\delta W^{(\text{applied})} = \sum_{i=1}^{N} \mathbf{F}_i^{(\text{applied})} \cdot \delta \mathbf{r}_i$$

plus the work done by the inertial forces

$$\delta W^{(\text{inertial})} = \sum_{i=1}^{N} (-m_i \ddot{\mathbf{r}}_i) \cdot \delta \mathbf{r}_i$$

in a virtual displacement $\delta \mathbf{r}_i$ is zero.

Let us first consider the work $\delta W^{(\text{applied})}$ done by the applied forces. If the 3N particle coordinates $\mathbf{r}_i$ are expressed in terms of a set of f generalized coordinates $q_a$,

$$\mathbf{r}_i = \mathbf{r}_i(q_1, q_2, \cdots, q_f; t),$$

the virtual displacements of the particles can be expressed in the form

$$\delta \mathbf{r}_i = \sum_{a=1}^{f} \frac{\partial \mathbf{r}_i}{\partial q_a} \delta q_a .$$

Note that there is no term in this equation coming from "variation in time." In a virtual displacement we imagine the system "frozen" at the time of interest and then imagine subjecting the system to displacements consistent with the constraints. Time is fixed in a virtual displacement. The work done by the applied forces can thus be written

$$\delta W^{(\text{applied})} = \sum_{a=1}^{f} \left( \sum_{i=1}^{N} \mathbf{F}_i^{(\text{applied})} \cdot \frac{\partial \mathbf{r}_i}{\partial q_a} \right) \delta q_a .$$

The expression in brackets,

$$Q_a = \sum_{i=1}^{N} \mathbf{F}_i^{(\text{applied})} \cdot \frac{\partial \mathbf{r}_i}{\partial q_a},$$

is called a **generalized force**. It may be thought of as the "component" of the force in the "direction" of the generalized coordinate $q_a$. Note, however, that $Q_a$ might not even have the dimensions of "force"; if, for example, the generalized coordinate is an angle, the generalized force is typically a torque. With this notation we can write the applied work in the form

$$\delta W^{(\text{applied})} = \sum_{a=1}^{f} Q_a \, \delta q_a .$$

This provides an alternate interpretation of generalized force as "applied work done per unit generalized displacement." If the applied forces are all conservative, they can be expressed as the negative gradients of the total potential energy $V(\mathbf{r}_1, \mathbf{r}_2, \cdots, \mathbf{r}_N; t)$ of the system,

$$\mathbf{F}_i^{(\text{applied})} = -\nabla_i V .$$

The generalized force is then given by

$$Q_a = -\frac{\partial V}{\partial q_a} .$$

In this case the work done by the applied forces is simply the negative of the change in the potential energy of the system,

$$\delta W^{(\text{applied})} = -\delta V .$$

Let us now consider the work $\delta W^{(\text{inertial})}$ done by the inertial forces. We shall see that this can be expressed in terms of the total kinetic energy

$$T = \sum_{i=1}^{N} \tfrac{1}{2} m_i |\dot{\mathbf{r}}_i|^2$$

of the system. Since the particle velocity can be written

$$\dot{\mathbf{r}}_i = \sum_{a=1}^{f} \frac{\partial \mathbf{r}_i}{\partial q_a} \dot{q}_a + \frac{\partial \mathbf{r}_i}{\partial t},$$

the kinetic energy is a function of the generalized coordinates $q_a$ and their first time derivatives, the **generalized velocities** $\dot{q}_a$. In particular, since the particle velocity is a linear function of the generalized velocities, the kinetic energy is a quadratic function of the generalized velocities. Now we have

$$\frac{\partial T}{\partial \dot{q}_a} = \sum_{i=1}^{N} m_i \dot{\mathbf{r}}_i \cdot \frac{\partial \dot{\mathbf{r}}_i}{\partial \dot{q}_a} = \sum_{i=1}^{N} m_i \dot{\mathbf{r}}_i \cdot \frac{\partial \mathbf{r}_i}{\partial q_a},$$

the second equality following from the preceding equation, which implies that

$$\frac{\partial \dot{\mathbf{r}}_i}{\partial \dot{q}_a} = \frac{\partial \mathbf{r}_i}{\partial q_a}.$$

In terms of the cartesian coordinates $(x,y,z)$ the kinetic energy for a single particle is $T = \frac{1}{2}m(\dot{x}^2 + \dot{y}^2 + \dot{z}^2)$, and the quantities $\partial T/\partial \dot{x} = m\dot{x}$, $\partial T/\partial \dot{y} = m\dot{y}$, $\partial T/\partial \dot{z} = m\dot{z}$ are the cartesian components of the linear momentum of the particle. The quantity $\partial T/\partial \dot{q}_a$ can thus be thought of as a **generalized momentum**,[1] the "component" of momentum in the "direction" of the generalized coordinate $q_a$ (compare the definition of generalized force). In view of Newton's second law it makes sense to consider its (total) time derivative

$$\frac{d}{dt}\left(\frac{\partial T}{\partial \dot{q}_a}\right) = \sum_{i=1}^{N} m_i \ddot{\mathbf{r}}_i \cdot \frac{\partial \mathbf{r}_i}{\partial q_a} + \sum_{i=1}^{N} m_i \dot{\mathbf{r}}_i \cdot \frac{d}{dt}\left(\frac{\partial \mathbf{r}_i}{\partial q_a}\right).$$

The first term on the right is exactly what we need for $\delta W^{(\text{inertial})}$. In the second term we can exchange the order of the time differentiation and the q-differentiation, since

$$\frac{d}{dt}\left(\frac{\partial \mathbf{r}_i}{\partial q_a}\right) = \sum_{b=1}^{f} \frac{\partial^2 \mathbf{r}_i}{\partial q_b \partial q_a}\dot{q}_b + \frac{\partial^2 \mathbf{r}_i}{\partial t \partial q_a} = \frac{\partial}{\partial q_a}\left(\sum_{b=1}^{f} \frac{\partial \mathbf{r}_i}{\partial q_b}\dot{q}_b + \frac{\partial \mathbf{r}_i}{\partial t}\right) = \frac{\partial}{\partial q_a}\left(\frac{d\mathbf{r}_i}{dt}\right).$$

The second term can thus be written in the form

$$\sum_{i=1}^{N} m_i \dot{\mathbf{r}}_i \cdot \frac{\partial \dot{\mathbf{r}}_i}{\partial q_a} = \frac{\partial}{\partial q_a} \sum_{i=1}^{N} \tfrac{1}{2} m_i |\dot{\mathbf{r}}_i|^2 = \frac{\partial T}{\partial q_a}.$$

The work done by the inertial forces in a virtual displacement finally becomes

$$\delta W^{(\text{inertial})} = \sum_{a=1}^{f}\sum_{i=1}^{N} (-m_i \ddot{\mathbf{r}}_i)\cdot \frac{\partial \mathbf{r}_i}{\partial q_a}\delta q_a = \sum_{a=1}^{f}\left(\frac{\partial T}{\partial q_a} - \frac{d}{dt}\left(\frac{\partial T}{\partial \dot{q}_a}\right)\right)\delta q_a.$$

Putting together our expressions for $\delta W^{(\text{applied})}$ and $\delta W^{(\text{inertial})}$, we see that d'Alembert's principle gives

---

[1]This definition will be modified on page 47 when we consider more general systems.

$$\sum_{a=1}^{f}\left(Q_a + \frac{\partial T}{\partial q_a} - \frac{d}{dt}\left(\frac{\partial T}{\partial \dot{q}_a}\right)\right)\delta q_a = 0.$$

Since the generalized coordinates $q_a$ are all independent, the coefficient of each of the virtual displacements $\delta q_a$ must vanish, which leads to the set of f equations

$$\frac{d}{dt}\left(\frac{\partial T}{\partial \dot{q}_a}\right) - \frac{\partial T}{\partial q_a} = Q_a \qquad a = 1,2,\cdots,f.$$

This set of f second order differential equations for the f generalized coordinates $q_a$ is known as **Lagrange's equations**. We have seen that if the applied forces are all conservative, the generalized force can be written

$$Q_a = -\frac{\partial V}{\partial q_a}$$

where V is the total potential energy. Since V does not depend on the velocities, Lagrange's equations become in this case

$$\frac{d}{dt}\left(\frac{\partial L}{\partial \dot{q}_a}\right) - \frac{\partial L}{\partial q_a} = 0 \qquad a = 1,2,\cdots,f$$

where

$$L = T - V,$$

the difference between the kinetic and potential energies, is called the **Lagrangian**.

Lagrange's equations provide one of the most convenient ways of writing down the equations of motion for a wide range of mechanical systems. We can proceed as follows:

1. Choose a set of generalized coordinates $(q_1, q_2, \ldots, q_f)$.
2. Express the kinetic energy T and potential energy V of the system in terms of these coordinates, their first time derivatives, and the time. Form the Lagrangian $L = T - V$.
3. Substitute L into Lagrange's equations and perform the indicated differentiations.

There are many advantages to such an approach. First, the unknown forces of constraint do not appear. Second, we can use *any* set of generalized coordinates to describe the configuration of the system, and in particular we can choose a set suited to the problem at hand; Lagrange's equations take the same general form no matter what set of coordinates we use. And finally, we need consider only *scalar* quantities, speed, kinetic and potential energy, as opposed to the *vector* quantities, acceleration, force, associated with Newtonian mechanics. Some examples will help make these features clear.

## Plane pendulum

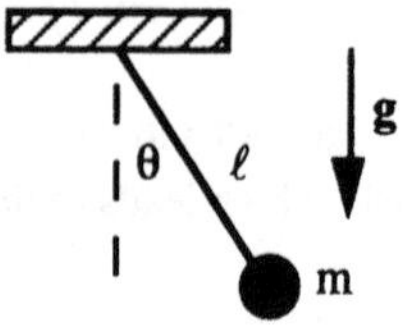


Fig. 3.01. Plane pendulum

A mass m is suspended from the ceiling by a string of length $\ell$ and swings back and forth in a vertical plane (Fig. 3.01). Gravity g acts vertically down. Let us use Lagrange's method to find the equation of motion. A suitable generalized coordinate is the angle $\theta$ the string makes with the vertical. In terms of this variable the kinetic energy is

$$T = \frac{1}{2}m\ell^2\dot{\theta}^2,$$

and the potential energy (relative to the ceiling) is

$$V = -mg\ell\cos\theta.$$

The Lagrangian is thus

$$L = \frac{1}{2}m\ell^2\dot{\theta}^2 + mg\ell\cos\theta.$$

Its derivative with respect to $\theta$ is

$$\frac{\partial L}{\partial \theta} = -mg\ell\sin\theta$$

and is physically the torque on the mass about the point of support. Its derivative with respect to $\dot{\theta}$, the generalized momentum associated with $\theta$, is

$$\frac{\partial L}{\partial \dot{\theta}} = m\ell^2\dot{\theta}$$

and is physically the angular momentum of the mass about the point of support. Lagrange's equation thus gives the equation of motion

$$m\ell^2\ddot{\theta} = -mg\ell\sin\theta.$$

We do not have to use the angle $\theta$ as generalized coordinate. We could, for example, instead use the horizontal displacement

$$x = \ell \sin\theta$$

of the mass from equilibrium. In terms of this variable the speed of the mass is

$$v = \frac{\dot{x}}{\cos\theta} = \frac{\ell\dot{x}}{\sqrt{\ell^2 - x^2}},$$

and the kinetic energy T and potential energy V are given by

$$T = \tfrac{1}{2}m\frac{\ell^2\dot{x}^2}{\ell^2 - x^2} \quad \text{and} \quad V = -mg\sqrt{\ell^2 - x^2}.$$

We form the Lagrangian $L = T - V$, substitute it into Lagrange's equation, perform the differentiations and simplify, and thus obtain the equation of motion in terms of the generalized coordinate x,

$$m\ddot{x} = -\frac{mx\dot{x}^2}{\ell^2 - x^2} - \frac{mgx}{\ell^2}\sqrt{\ell^2 - x^2}.$$

This equation for x is much more complicated looking than the equation for the angle $\theta$, although they both describe the same physical system. It makes sense to try to find generalized coordinates for which the equations of motion are "simple." At the moment we have no systematic way to do this, although experience helps. In Chapter VII we study in detail transformations from one set of variables to another. Indeed, in Chapter VIII we use this approach to solve the equations of motion, by finding a transformation to a set of variables for which the equations can be solved by inspection.

## Spherical pendulum

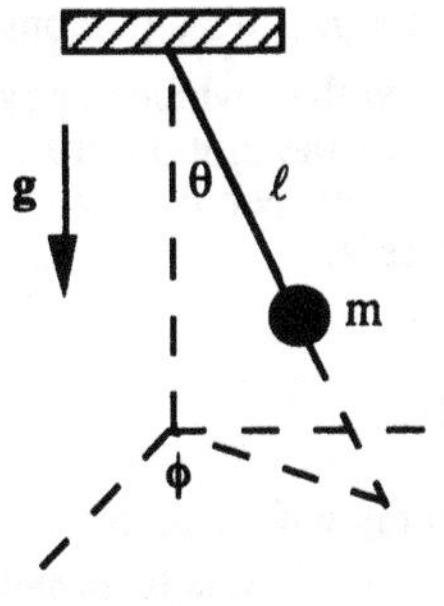


Fig. 3.02. Spherical pendulum

A mass m is suspended from the ceiling by a string of length $\ell$ as in the plane pendulum. Now, however, the mass is allowed to swing in all directions (Fig. 3.02). The situation is thus the same as a mass constrained to move on the surface of a sphere. Gravity g acts vertically down. Let us again use Lagrange's method to find the equations of motion. Suitable generalized coordinates are the spherical polar angles $\theta$ and $\phi$. In terms of these variables the kinetic energy is

$$T = \tfrac{1}{2} m\ell^2(\dot\theta^2 + \dot\phi^2 \sin^2\theta).$$

The potential energy is the same as for the plane pendulum, so the Lagrangian is

$$L = \tfrac{1}{2} m\ell^2(\dot\theta^2 + \dot\phi^2 \sin^2\theta) + mg\ell\cos\theta.$$

Its derivatives with respect to the coordinates $\theta$ and $\phi$ and with respect to the generalized velocities $\dot\theta$ and $\dot\phi$ are

$$\frac{\partial L}{\partial \theta} = m\ell^2 \sin\theta\cos\theta\dot\phi^2 - mg\ell\sin\theta \qquad \frac{\partial L}{\partial \phi} = 0$$

$$\frac{\partial L}{\partial \dot\theta} = m\ell^2\dot\theta \qquad \frac{\partial L}{\partial \dot\phi} = m\ell^2 \sin^2\theta\dot\phi\,,$$

so Lagrange's equations of motion are

$$m\ell^2\ddot\theta = m\ell^2 \sin\theta\cos\theta\dot\phi^2 - mg\ell\sin\theta$$

$$\frac{d}{dt}\left(m\ell^2 \sin^2\theta\dot\phi\right) = 0\ .$$

In this example we notice that the Lagrangian does not contain the generalized coordinate $\phi$ (such coordinates are often called **cyclic coordinates**). Physically, the reason for this is that the system remains unchanged under rotations about a vertical axis; the system is **invariant** under these rotations. Lagrange's equations then imply that the generalized momentum $m\ell^2 \sin^2\theta\dot\phi$ associated with $\phi$, which is physically the angular momentum in the vertical direction, is a constant of the motion; the angular momentum is **conserved**. This illustrates a general connection between invariance properties and conservation laws which we explore in detail in Chapter V.

## Electromagnetic interaction[2]

There are, of course, systems whose forces are not obtainable from a potential $V(\mathbf{r},t)$. For some of these it may nevertheless be possible to write down a Lagrangian, a function of the q's, $\dot q$'s, and t, which when substituted into Lagrange's equations gives the correct equations of motion. One might wonder why we need a Lagrangian if we already

[2]John David Jackson, *Classical Electrodynamics*, (John Wiley and Sons, New York, NY, 1962; 1975), 2nd ed., particularly sects. 6.4, 6.5, 12.1.

know the equations of motion. A reason is that by describing the system in terms of a Lagrángian, we can then bring to bear on the problem all the ideas and techniques of advanced mechanics . Also, in many cases we do *not* know the equations of motion *a priori*. We may then try to find suitable equations by writing down a suitable Lagrangian, using general principles and known cases as a guide.

An important case is the interaction of a particle of charge e with an electromagnetic field. If the field is described by scalar and vector potentials $\phi$ and $\mathbf{A}$, so that the electric and magnetic fields $\mathbf{E}$ and $\mathbf{B}$ are given by

$$\mathbf{E} = -\nabla\phi - \frac{1}{c}\frac{\partial \mathbf{A}}{\partial t} \quad \text{and} \quad \mathbf{B} = \nabla \times \mathbf{A},$$

we can show that a suitable Lagrangian is

$$L = \tfrac{1}{2}m|\dot{\mathbf{r}}|^2 - e\phi + (e/c)\dot{\mathbf{r}}\cdot\mathbf{A}.$$

Since Lagrangians are no longer necessarily the difference between the kinetic and potential energies, it is convenient to change, from now on, our definition of the **generalized momentum** to $p_a = \partial L/\partial \dot{q}_a$ from our earlier expression $\partial T/\partial \dot{q}_a$ (the two expressions agree if $L(q,\dot{q},t) = T(q,\dot{q},t) - V(q,t)$). The generalized momentum associated with $\mathbf{r}$ is then

$$\mathbf{p} = \frac{\partial L}{\partial \dot{\mathbf{r}}} = m\dot{\mathbf{r}} + \frac{e}{c}\mathbf{A}$$

and is the sum of the ordinary "kinetic momentum" $m\dot{\mathbf{r}}$ and a term $(e/c)\mathbf{A}$ which we shall consider in more detail in the next section. We shall see that it is "field momentum." The time derivative of the generalized momentum is

$$\frac{d}{dt}\left(\frac{\partial L}{\partial \dot{\mathbf{r}}}\right) = m\ddot{\mathbf{r}} + \frac{e}{c}\frac{\partial \mathbf{A}}{\partial t} + \frac{e}{c}(\dot{\mathbf{r}}\cdot\nabla)\mathbf{A}.$$

The derivative of the Lagrangian with respect to the generalized coordinate $\mathbf{r}$ is

$$\nabla L = -e\nabla\phi + (e/c)\nabla(\dot{\mathbf{r}}\cdot\mathbf{A}),$$

so Lagrange's equations give

$$m\ddot{\mathbf{r}} = e\left(-\nabla\phi - \frac{1}{c}\frac{\partial \mathbf{A}}{\partial t}\right) + \frac{e}{c}\left[\nabla(\dot{\mathbf{r}}\cdot\mathbf{A}) - (\dot{\mathbf{r}}\cdot\nabla)\mathbf{A}\right].$$

The term in round brackets on the right is the electric field $\mathbf{E}$. A vector identity shows that the term in square brackets can be written

$$\nabla(\dot{\mathbf{r}}\cdot\mathbf{A}) - (\dot{\mathbf{r}}\cdot\nabla)\mathbf{A} = \dot{\mathbf{r}}\times(\nabla\times\mathbf{A}) = \dot{\mathbf{r}}\times\mathbf{B}$$

where **B** is the magnetic field. Lagrange's equations become

$$m\ddot{\mathbf{r}} = e\mathbf{E} + \frac{e}{c}\dot{\mathbf{r}} \times \mathbf{B}.$$

We recognize the right-hand side as the electromagnetic force, the **Lorentz force**, which an electromagnetic field $(\mathbf{E}, \mathbf{B})$ exerts on a particle of charge e. This shows that the stated Lagrangian is indeed "suitable."

The electromagnetic potentials $(\phi, \mathbf{A})$ for a given system are not unique. We can always subject them to a **gauge transformation**, the old potentials $(\phi, \mathbf{A})$ being replaced by new potentials $(\phi', \mathbf{A}')$ with

$$\phi' = \phi - (1/c)\,\partial\lambda/\partial t$$
$$\mathbf{A}' = \mathbf{A} + \nabla\lambda$$

where $\lambda$ is an arbitrary (single-valued) function of space and time, without affecting any of the physics. In particular, under such a transformation the electromagnetic fields $(\mathbf{E}, \mathbf{B})$ remain unchanged. Since the Lagrangian contains the potentials rather than the fields, we might worry that it might lead to non-gauge-invariant physics. We can see, of course, that the resulting equation of motion *is* gauge invariant, but it would be nice to see this from the Lagrangian. Under the above gauge transformation, the Lagrangian L is replaced by a new Lagrangian $L'$, with

$$L' = L + \frac{e}{c}\frac{d\lambda}{dt}.$$

The two Lagrangians differ by the total time derivative of $(e/c)\lambda$. Now it is easy to show that such a term gives zero identically when substituted into Lagrange's equations, so the two Lagrangians are in fact equivalent. We should keep in mind that this applies to any two Lagrangians which differ by any total time derivative. It is used again in Chapter V.

We often wish to describe the motion of electrons and other subatomic particles in electric and magnetic fields, and for such particles speeds v close to the speed c of light are common. This means that relativistic mechanics[3] should be used, the left-hand side of the above equation of motion being replaced by $\frac{d}{dt}\frac{m\mathbf{v}}{\sqrt{1-(v/c)^2}}$. The resulting equation can be obtained from a Lagrangian

$$L = -mc^2\sqrt{1-(v/c)^2} - e\phi + (e/c)\mathbf{v}\cdot\mathbf{A}$$

with modified first or "free particle" term. This term is not the (relativistic) "kinetic energy," and the second and third interaction terms are not "potential energy," but no

[3]See, for example, Wolfgang Rindler, *Introduction to Special Relativity*, (Oxford University Press, Oxford, UK, 1982), Chap. V.

matter; when this L is substituted into Lagrange's equations, the correct equations of motion result, and this is all we require.

We shall see in the next chapter that the **action** $S = \int L\,dt$, the time integral of the Lagrangian L, is a more fundamental physical quantity than the Lagrangian itself. The "free particle term" in the relativistic Lagrangian gives a contribution to the element of action

$$-mc^2\sqrt{1-(v/c)^2}\,dt = -mc^2\,d\tau$$

where $d\tau$ is the element of **proper time** (the time read by a clock carried by the particle), a relativistic scalar. If the stated Lagrangian is to describe relativistically invariant physics, the remaining interaction terms

$$(-e\phi + (e/c)\mathbf{v}\cdot\mathbf{A})\,dt = -(e/c)(\phi c dt - \mathbf{A}\cdot d\mathbf{r})$$

should also form a relativistic scalar. And indeed they do, being just the scalar product $-(e/c)A\cdot dx$ of the relativistic four-vectors $A = (\phi, \mathbf{A})$ and $dx = (cdt, d\mathbf{r})$. We can turn this argument around and use it to suggest the form of the interaction, the interaction terms clearly being a relativistically invariant extension of $-V\,dt$ where $V = e\phi$ is the potential energy of a charge in an electrostatic field.

## Interaction of an electric charge and a magnet[4]

In order to further our understanding of the electromagnetic interaction, let us consider the interaction of a point electric charge e and a small magnet with magnetic dipole moment $\mathbf{m}$. We restrict our attention to the quasi-static limit in which radiation is neglected, and the charge and magnet and their quasi-static fields form a closed system. This system would seem to be trivially simple, but as we shall see, it has some surprises in

---

[4]Most electromagnetism texts do not discuss the aspects of this topic with which we are concerned here, but see:

W. Shockley and R. P. James, "'Try Simplest Cases' Discovery of 'Hidden Momentum' Forces on 'Magnetic Currents'," Phys. Rev. Lett. **18**, 876-879 (1967).

W. Shockley, "'Hidden Linear Momentum' Related to the α·E Terms for a Dirac-Electron Wave Packet in an Electric Field," Phys. Rev. Lett. **20**, 343-346 (1968).

P. Penfield and H. Haus, *The Electrodynamics of Moving Media* (MIT, Cambridge, MA, 1967), p. 215; "Force on a Current Loop," Phys. Lett. **26A**, 412-413 (1968).

Sidney Coleman and J. H. VanVleck, "Origin of 'Hidden Momentum Forces' on Magnets," Phys. Rev. **171**, 1370-1375 (1968).

W. H. Furry, "Examples of Momentum Distributions in the Electromagnetic Field and in Matter," Am. J. Phys. **37**, 621-636 (1969).

M. G. Calkin, "Linear Momentum of Quasistatic Electromagnetic Fields," Am. J. Phys. **34**, 921-925 (1966); "Linear Momentum of the Source of a Static Electromagnetic Field," Am. J. Phys. **39**, 513-516 (1971).

Y. Aharonov, P. Pearle, and L. Vaidman, "Comment on 'Proposed Aharonov-Casher effect: Another example of an Aharonov-Bohm effect arising from a classical lag'," Phys. Rev. A **37**, 4052-4055 (1988).

Lev Vaidman, "Torque and force on a magnetic dipole," Am. J. Phys. **58**, 978-983 (1990).

store. Choose a coordinate system in which the magnet is instantaneously at rest at $\mathbf{r}_m$, and the charge is at $\mathbf{r}_e$ and has velocity $\mathbf{v}_e = d\mathbf{r}_e/dt$ (Fig. 3.03).

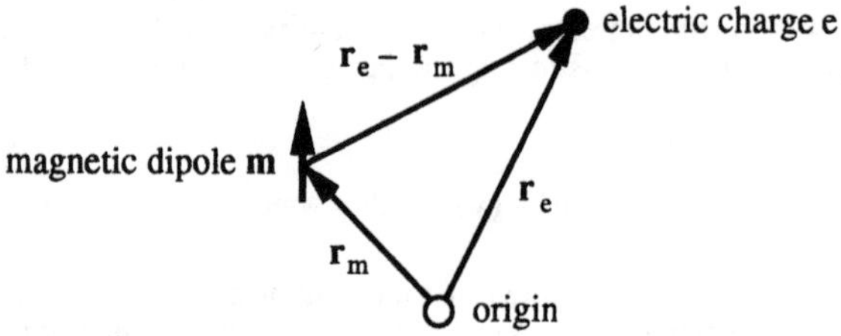


Fig. 3.03. Electric charge and magnet

The magnet generates a magnetic field

$$\mathbf{B}_m(\mathbf{r}) = \nabla \times \mathbf{A}_m(\mathbf{r})$$

where

$$\mathbf{A}_m(\mathbf{r}) = \frac{\mathbf{m} \times (\mathbf{r} - \mathbf{r}_m)}{|\mathbf{r} - \mathbf{r}_m|^3}$$

is the vector potential in the gauge in which $\nabla \cdot \mathbf{A}_m = 0$; it is the transverse vector potential. This magnetic field exerts a Lorentz force

$$\mathbf{F}_{\text{mone}} = \frac{e}{c}\mathbf{v}_e \times \mathbf{B}_m(\mathbf{r}_e)$$
$$= \frac{e}{c}\nabla_e(\mathbf{v}_e \cdot \mathbf{A}_m(\mathbf{r}_e)) - \frac{e}{c}\frac{d\mathbf{A}_m(\mathbf{r}_e)}{dt}$$

on the moving charge. The second line, expressing the force in terms of the vector potential, follows from reversing some of the steps in the previous section. On the other hand, the moving electric charge generates a magnetic field

$$\mathbf{B}_e(\mathbf{r}) = \frac{1}{c}\mathbf{v}_e \times \mathbf{E}_e(\mathbf{r})$$

where

$$\mathbf{E}_e(\mathbf{r}) = e\frac{\mathbf{r} - \mathbf{r}_e}{|\mathbf{r} - \mathbf{r}_e|^3}$$

is the electric field due to the charge. This magnetic field exerts a Lorentz force[5]

$$\mathbf{F}_{\mathrm{eonm}} = \nabla_{\mathrm{m}}(\mathbf{m}\cdot\mathbf{B}_{\mathrm{e}}(\mathbf{r}_{\mathrm{m}}))$$

on the magnet. Now we have

$$\mathbf{m}\cdot\mathbf{B}_{\mathrm{e}}(\mathbf{r}_{\mathrm{m}}) = \frac{1}{c}\mathbf{m}\cdot\mathbf{v}_{\mathrm{e}}\times\mathbf{E}_{\mathrm{e}}(\mathbf{r}_{\mathrm{m}}) = \frac{e}{c}\mathbf{v}_{\mathrm{e}}\cdot\frac{\mathbf{m}\times(\mathbf{r}_{\mathrm{e}}-\mathbf{r}_{\mathrm{m}})}{|\mathbf{r}_{\mathrm{e}}-\mathbf{r}_{\mathrm{m}}|^3} = \frac{e}{c}\mathbf{v}_{\mathrm{e}}\cdot\mathbf{A}_{\mathrm{m}}(\mathbf{r}_{\mathrm{e}}).$$

From this and from $\nabla_{\mathrm{m}} = -\nabla_{\mathrm{e}}$ (the arguments depending only on $\mathbf{r}_{\mathrm{e}} - \mathbf{r}_{\mathrm{m}}$), we see that $\mathbf{F}_{\mathrm{eonm}}$ is the negative of the *first* term in $\mathbf{F}_{\mathrm{mone}}$, and thus

$$\mathbf{F}_{\mathrm{mone}} + \mathbf{F}_{\mathrm{eonm}} = -\frac{e}{c}\frac{d\mathbf{A}_{\mathrm{m}}(\mathbf{r}_{\mathrm{e}})}{dt}.$$

The electromagnetic forces which the magnet exerts on the charge and which the charge exerts on the magnet are not equal and opposite; Newton's third law does not hold, even in this quasi-static limit. This is the first surprise. If we use Newton's second law to set

$$\mathbf{F}_{\mathrm{mone}} = \frac{d\mathbf{p}_{\mathrm{e}}}{dt} \quad \text{and} \quad \mathbf{F}_{\mathrm{eonm}} = \frac{d\mathbf{p}_{\mathrm{m}}}{dt}$$

where $\mathbf{p}_{\mathrm{e}}$ and $\mathbf{p}_{\mathrm{m}}$ are the momenta of the charge and of the magnet, we see that the total mechanical momentum $\mathbf{p}_{\mathrm{e}} + \mathbf{p}_{\mathrm{m}}$ is *not* constant. However, we *do* have

$$\mathbf{p}_{\mathrm{total}} \equiv \mathbf{p}_{\mathrm{e}} + \mathbf{p}_{\mathrm{m}} + (e/c)\mathbf{A}_{\mathrm{m}}(\mathbf{r}_{\mathrm{e}}) = \text{constant}.$$

The third term $(e/c)\mathbf{A}_{\mathrm{m}}(\mathbf{r}_{\mathrm{e}})$ is the momentum in the electromagnetic field which consists of the electric field of the charge and the magnetic field of the magnet. The usual expression for electromagnetic field momentum is the integral over all space of a field momentum density $(1/4\pi c)\mathbf{E}\times\mathbf{B}$. One can show, however, that in the quasi-static limit

$$\mathbf{p}_{\mathrm{em}} = \frac{1}{4\pi c}\int \mathbf{E}\times\mathbf{B}\,d^3\mathbf{r} = \frac{1}{c}\int \rho\mathbf{A}\,d^3\mathbf{r}$$

where $\rho$ is the electric charge density and $\mathbf{A}$ is the transverse vector potential. Our expression $(e/c)\mathbf{A}_{\mathrm{m}}(\mathbf{r}_{\mathrm{e}})$ for the field momentum is thus equivalent to the usual expression. To summarize: we must include electromagnetic field momentum if we want the momentum of the (closed) charge-magnet system to remain constant.

Now comes the second surprise. The momentum of the charge is $M_{\mathrm{e}}\mathbf{v}_{\mathrm{e}}$, but the momentum of the magnet is *not* $M_{\mathrm{m}}\mathbf{v}_{\mathrm{m}}$. One can show that the total momentum of any static system is zero. Thus, even if the charge and the magnet are at rest, there must still be

---

[5]John David Jackson, *Classical Electrodynamics*, (John Wiley and Sons, New York, NY, 1975), 2nd ed., p. 185.

mechanical momentum somewhere to cancel the non-zero field momentum. The only place this can be, the only place where there is moving matter, is in the magnet, in the circulating charges which generate the magnetic moment. The magnet at rest thus has net "hidden" mechanical momentum $-(e/c)\mathbf{A}_m(\mathbf{r}_e)$. If the magnet moves with low velocity $\mathbf{v}_m$, it has additional momentum $M_m\mathbf{v}_m$ for a net momentum

$$\mathbf{p}_m = M_m\mathbf{v}_m - (e/c)\mathbf{A}_m(\mathbf{r}_e).$$

The total momentum of the charge-magnet system, mechanical plus electromagnetic, is thus

$$\mathbf{p}_{total} = M_e\mathbf{v}_e + [M_m\mathbf{v}_m - (e/c)\mathbf{A}_m(\mathbf{r}_e)] + (e/c)\mathbf{A}_m(\mathbf{r}_e) = M_e\mathbf{v}_e + M_m\mathbf{v}_m$$

and is constant. The last expression looks just like the ordinary mechanical momentum of two particles, with no suggestion of field momentum or of hidden mechanical momentum, these latter contributions canceling. We have $M_e(d\mathbf{v}_e/dt) = -M_m(d\mathbf{v}_m/dt)$, so if we define "force" as "mass times acceleration" rather than as "rate of change of momentum," the "force" the magnet exerts on the charge *is* equal and opposite the "force" the charge exerts on the magnet. In this form Newton's third law holds.

To see the origin of the hidden mechanical momentum in the magnet, let us consider a simple model. Suppose that the magnetic moment is due to a particle of mass m and charge q circulating around a closed loop (Fig. 3.04).

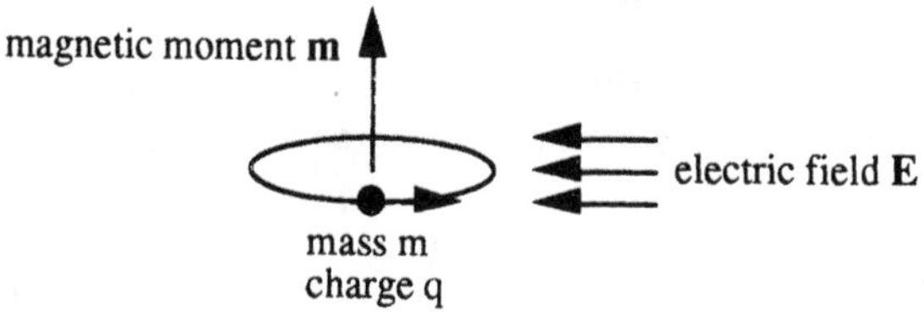


Fig. 3.04. Magnetic moment in an electric field

The average (relativistic) momentum of such a system is

$$\langle\mathbf{p}\rangle = \frac{1}{\tau}\oint\frac{m\mathbf{v}}{\sqrt{1-(v/c)^2}}dt = \frac{1}{\tau}\oint\frac{m\,d\mathbf{l}}{\sqrt{1-(v/c)^2}}$$

where $\tau$ is the time to go around. If the speed of the particle is constant, we have $\langle\mathbf{p}\rangle \propto \oint d\mathbf{l} = 0$ as expected. Now suppose that the loop is in an electric field. The speed of the particle is then not constant, but varies so as to keep the (relativistic) energy

$$\varepsilon = \frac{mc^2}{\sqrt{1-(v/c)^2}} + q\phi,$$

where $\phi$ is the electrostatic potential, constant. The average momentum becomes

$$\langle \mathbf{p} \rangle = \frac{1}{c^2 \tau} \oint (\varepsilon - q\phi)\, dl = -\frac{I}{c^2} \oint \phi\, dl$$

where $I = q/\tau$ is the average electric current around the loop. A vector identity enables us to rewrite this expression for the momentum as

$$\langle \mathbf{p} \rangle = -\frac{I}{c^2} \int d\mathbf{a} \times \nabla\phi = \frac{I}{c^2} \int d\mathbf{a} \times \mathbf{E},$$

the integration now extending over a surface which has the loop as its edge. In the situations in which we are interested the electric field is essentially constant over the area of the loop, and we can factor it out, obtaining finally

$$\langle \mathbf{p} \rangle = \frac{1}{c} \mathbf{m} \times \mathbf{E}$$

where $\mathbf{m} = (I/c)\int d\mathbf{a}$ is the magnetic dipole moment. It is easy to see that this expression for $\langle \mathbf{p} \rangle$ is, for our system, the same as $-(e/c)\mathbf{A}_m(\mathbf{r}_e)$.

Because of these considerations, the equation of motion of a small magnet of mass M and magnetic moment $\mathbf{m}$ in an electromagnetic field $(\mathbf{E}, \mathbf{B})$ is (now dropping subscripts)

$$\frac{d}{dt}\left( M\mathbf{v} + \frac{1}{c} \mathbf{m} \times \mathbf{E} \right) = \nabla(\mathbf{m} \cdot \mathbf{B})$$

in the instantaneous rest frame of the magnet. Carrying out the differentiations, we find

$$M \frac{d\mathbf{v}}{dt} + \frac{1}{c} \frac{d\mathbf{m}}{dt} \times \mathbf{E} + \frac{1}{c} \mathbf{m} \times \frac{\partial \mathbf{E}}{\partial t} = (\mathbf{m} \cdot \nabla)\mathbf{B} + \mathbf{m} \times (\nabla \times \mathbf{B}).$$

The last terms on both sides can be simplified if we make use of the Maxwell equation

$$\nabla \times \mathbf{B} = \frac{4\pi}{c} \mathbf{J} + \frac{1}{c} \frac{\partial \mathbf{E}}{\partial t}$$

where $\mathbf{J}$ is the electric current density of the source of the field at the location of the magnet. We thus find

$$M \frac{d\mathbf{v}}{dt} = (\mathbf{m} \cdot \nabla)\mathbf{B} - \frac{1}{c} \frac{d\mathbf{m}}{dt} \times \mathbf{E} + \frac{4\pi}{c} \mathbf{m} \times \mathbf{J}.$$

The first two terms on the right are what we would get from a picture of a magnetic dipole which consisted of positive and negative magnetic poles separated by a small distance (as for an electric dipole). The presence or absence of the last term $(4\pi/c)\mathbf{m} \times \mathbf{J}$ in the equation of motion thus enables us to determine whether the magnetic dipole moment is due to circulating electric charges or to magnetic poles. For example, experiments show[6] that this term is of importance for a neutron moving through magnetic material; the neutron thus behaves as if its magnetic moment were due to circulating electric charges rather than to magnetic poles.

## Exercises

**Do exercises** 3 to 10 from Chapter II using Lagrangian methods.

1.

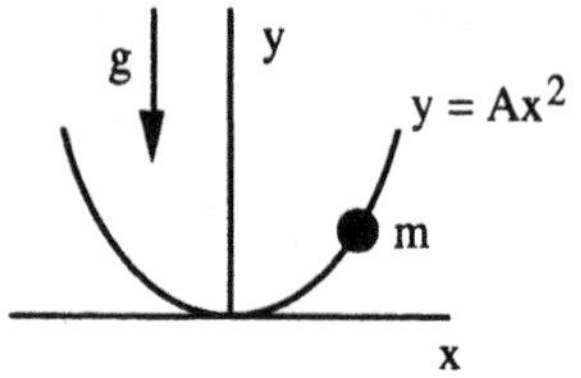


A bead of mass m slides without friction along a wire which has the shape of a parabola $y = Ax^2$ with axis vertical in the earth's gravitational field g.
(a) Find the Lagrangian, taking as generalized coordinate the horizontal displacement x.
(b) Write down Lagrange's equation of motion.

2.

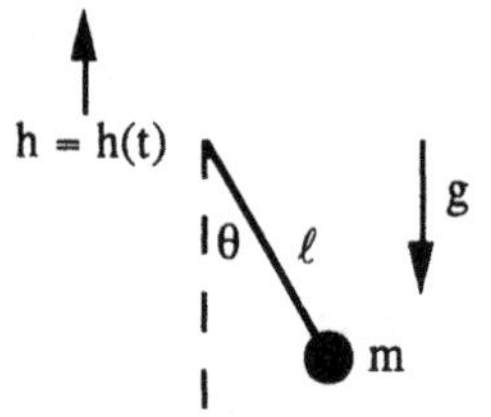


The point of support of a simple plane pendulum moves vertically according to $y = h(t)$, where $h(t)$ is some given function of time.

(a) Find the Lagrangian, taking as generalized coordinate the angle θ the pendulum makes with the vertical.
(b) Write down Lagrange's equation of motion, showing in particular that the pendulum behaves like a simple pendulum in a gravitational field $g + \ddot{h}$.

[6] C. G. Shull, E. O. Wollan, and W. A. Strauser, "Magnetic Structure of Magnetite and Its Use in Studying the Neutron Magnetic Interaction," Phys. Rev. **81**, 483-484 (1951); D. J. Hughes and M. J. Burgy, "Reflection of Neutrons from Magnetized Mirrors," Phys. Rev. **81**, 498-506 (1951).

3.

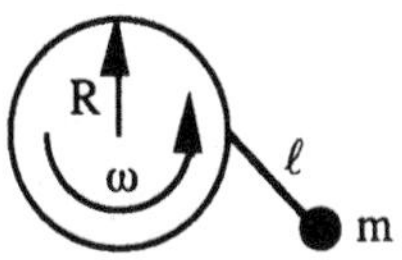


A mass m is attached to one end of a light rod of length $\ell$. The other end of the rod is pivoted so that the rod can swing in a plane. The pivot rotates in the same plane at angular velocity $\omega$ in a circle of radius R. Show that this "pendulum" behaves like a simple pendulum in a gravitational field $g = \omega^2 R$ for all values of $\ell$ and all amplitudes of oscillation.

4.

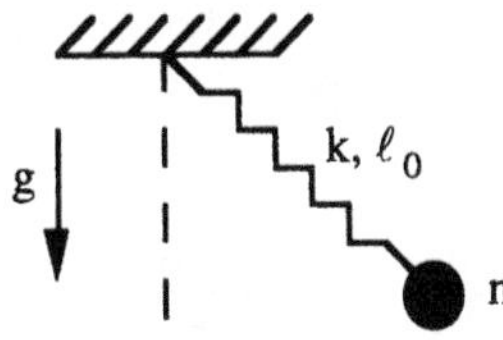


A pendulum is formed by suspending a mass m from the ceiling, using a spring of unstretched length $\ell_0$ and spring constant k.

(a) Choose, and show on a diagram, appropriate generalized coordinates, assuming that the pendulum moves in a fixed vertical plane.

(b) Set up the Lagrangian using your generalized coordinates.

(c) Write down the explicit Lagrange's equations of motion for your generalized coordinates.

5.

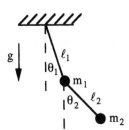

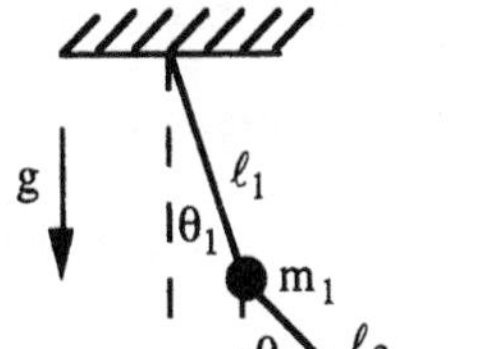


A double plane pendulum consists of two simple pendulums, with one pendulum suspended from the bob of the other. The "upper" pendulum has mass $m_1$ and length $\ell_1$, the "lower" pendulum has mass $m_2$ and length $\ell_2$, and both pendulums move in the same vertical plane.

(a) Find the Lagrangian, using as generalized coordinates the angles $\theta_1$ and $\theta_2$ the pendulums make with the vertical.

(b) Write down Lagrange's equations of motion.

6.

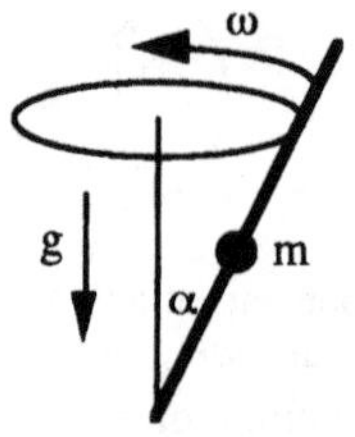


A bead of mass m slides on a long straight wire which makes an angle $\alpha$ with, and rotates with constant angular velocity $\omega$ about, the upward vertical. Gravity g acts vertically downwards.
(a) Choose an appropriate generalized coordinate and find the Lagrangian.
(b) Write down the explicit Lagrange's equation of motion.

7.

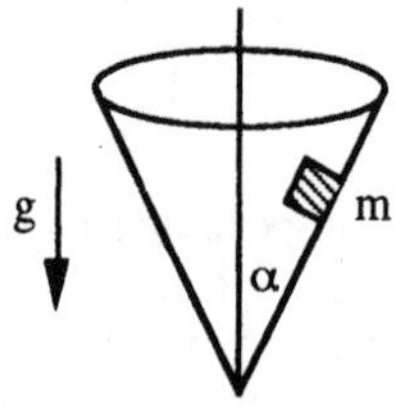


A particle of mass m slides on the inner surface of a cone of half angle $\alpha$. The axis of the cone is vertical with vertex downward. Gravity g acts vertically downwards.
(a) Choose and show on a diagram suitable generalized coordinates, and find the Lagrangian.
(b) Write down the explicit equations of motion for your generalized coordinates.

8. Using **spherical polar coordinates** $(r,\theta,\phi)$ defined by

$$x = r\sin\theta\cos\phi \quad y = r\sin\theta\sin\phi \quad z = r\cos\theta,$$

write down the Lagrangian and find the explicit Lagrange's equations of motion for a particle of mass m moving in a central potential $V(r)$.

9. For some problems **paraboloidal coordinates** $(\xi,\eta,\phi)$ defined by

$$x = \xi\eta\cos\phi \quad y = \xi\eta\sin\phi \quad z = \tfrac{1}{2}(\xi^2 - \eta^2)$$

turn out to be convenient.
(a) Show that the surfaces $\xi$ = const. or $\eta$ = const. are paraboloids of revolution about the z-axis with focus at the origin and semi-latus-rectum $\xi^2$ or $\eta^2$.
(b) Express the kinetic energy of a particle of mass m in terms of paraboloidal coordinates and their first time derivatives.
(Ans. $T = \frac{1}{2}m\left(\xi^2 + \eta^2\right)\left(\dot{\xi}^2 + \dot{\eta}^2\right) + \frac{1}{2}m\xi^2\eta^2\dot{\phi}^2$)

10. The motion of a particle of mass m is given by Lagrange's equations with Lagrangian

$$L = \exp(\alpha t/m)(T - V)$$

where $\alpha$ is a constant, $T = \frac{1}{2}m(\dot{x}^2 + \dot{y}^2 + \dot{z}^2)$ is the kinetic energy, and $V = V(x,y,z)$ is the potential energy. Write down the equations of motion and interpret.

11. A system with two degrees of freedom $(x,y)$ is described by a Lagrangian

$$L = \tfrac{1}{2}m(a\dot{x}^2 + 2b\dot{x}\dot{y} + c\dot{y}^2) - \tfrac{1}{2}k(ax^2 + 2bxy + cy^2)$$

where a, b, and c are constants, with $b^2 \neq ac$. Write down Lagrange's equations of motion and thereby identify the system. Consider in particular the cases $a = c = 0,\ b \neq 0$ and $a = -c,\ b = 0$.

12. The Lagrangian for two particles of masses $m_1$ and $m_2$ and coordinates $\mathbf{r}_1$ and $\mathbf{r}_2$, interacting via a potential $V(\mathbf{r}_1 - \mathbf{r}_2)$, is

$$L = \tfrac{1}{2}m_1|\dot{\mathbf{r}}_1|^2 + \tfrac{1}{2}m_2|\dot{\mathbf{r}}_2|^2 - V(\mathbf{r}_1 - \mathbf{r}_2).$$

(a) Rewrite the Lagrangian in terms of the **center of mass coordinates** $\mathbf{R} = \dfrac{m_1\mathbf{r}_1 + m_2\mathbf{r}_2}{m_1 + m_2}$ and **relative coordinates** $\mathbf{r} = \mathbf{r}_1 - \mathbf{r}_2$.

(b) Use Lagrange's equations to show that the center of mass and relative motions separate, the center of mass moving with constant velocity, and the relative motion being like that of a particle of **reduced mass** $\dfrac{m_1 m_2}{m_1 + m_2}$ in a potential $V(\mathbf{r})$.

13. Consider the motion of a free particle, with Lagrangian

$$L = \tfrac{1}{2}m(\dot{x}^2 + \dot{y}^2 + \dot{z}^2),$$

as viewed from a rotating coordinate system

$$x' = x\cos\theta + y\sin\theta, \quad y' = -x\sin\theta + y\cos\theta, \quad z' = z$$

where the angle $\theta = \theta(t)$ is some given function of time.

(a) Show that in terms of these coordinates the Lagrangian takes the form

$$L = \tfrac{1}{2}m[(\dot{x}'^2 + \dot{y}'^2 + \dot{z}'^2) + 2\omega(x'\dot{y}' - y'\dot{x}') + \omega^2(x'^2 + y'^2)]$$

where $\omega = d\theta/dt$ is the angular velocity.

(b) Write down Lagrange's equations of motion, and show that they look like those for a particle which is acted on by a "force." The part of the "force" proportional to $\omega$ is called the **Coriolis force**, that proportional to $\omega^2$ is called the **centrifugal force**, and that proportional to $d\omega/dt$ is called the **Euler force**. Identify the components of these "forces."

14. (a) Write down the equations of motion resulting from a Lagrangian

$$L = \tfrac{1}{2}m(\dot{x}^2 + \dot{y}^2 + \dot{z}^2) - V(r) + (eB/2c)(x\dot{y} - y\dot{x}),$$

and show that they are those for a particle of mass m and charge e moving in a central potential $V(r)$ together with a uniform magnetic field B which points in the z-direction.

(b) Suppose, instead of the inertial cartesian coordinate system $(x,y,z)$, we use a rotating system $(x',y',z')$ with

$$x' = x\cos\omega t + y\sin\omega t, \quad y' = -x\sin\omega t + y\cos\omega t, \quad z' = z.$$

Change variables, obtaining the above Lagrangian in terms of $(x',y',z')$ and their first time derivatives. Show that we can eliminate the term linear in B by an appropriate choice of $\omega$ (this is **Larmor's theorem**: the effect of a weak magnetic field on a system is to induce a uniform rotation at frequency $\omega_L$, the Larmor frequency).

15. Show that the equations of motion of an electric charge e interacting with a magnet of moment **m** can be obtained from a Lagrangian

$$L = \tfrac{1}{2}M_e v_e^2 + \tfrac{1}{2}M_m v_m^2 + (e/c)(\mathbf{v}_e - \mathbf{v}_m)\cdot\mathbf{A}(\mathbf{r}_e - \mathbf{r}_m),$$

where

$$\mathbf{A}(\mathbf{r}_e - \mathbf{r}_m) = \frac{\mathbf{m}\times(\mathbf{r}_e - \mathbf{r}_m)}{|\mathbf{r}_e - \mathbf{r}_m|^3}$$

is the vector potential at the charge due to the magnet.
(Y. Aharonov and A. Casher, "Topological Quantum Effects for Neutral Particles," Phys. Rev. Lett. **53**, 319-321 (1984)).

# CHAPTER IV

# THE PRINCIPLE OF STATIONARY ACTION OR HAMILTON'S PRINCIPLE

The dynamical behavior of a mechanical system has to this point been described by differential equations, by Newton's laws or, more generally, by Lagrange's equations. In this chapter we consider a different way of describing the dynamics. We show that a mechanical system moves from one configuration to another in such a way as to make a certain integral over the motion, called the action integral, stationary. Apart from the intrinsic philosophical interest of this principle, the ideas and techniques we develop have applications in other areas of physics, some of which we also discuss.

## Principle of stationary action

The configuration of a system with f degrees of freedom is specified by the values of a set of f generalized coordinates $q_a$. These coordinates form an f dimensional cartesian space called **configuration space**. Each configuration of the system corresponds to a point in this space. As the configuration of the system and hence the coordinates $q_a$ change with time, this point moves. A convenient way to represent this motion is by a path in an $f+1$ dimensional space, sometimes called **extended configuration space**, consisting of the f generalized coordinates and the time (Fig. 4.01). This chapter is concerned with the properties of such paths.

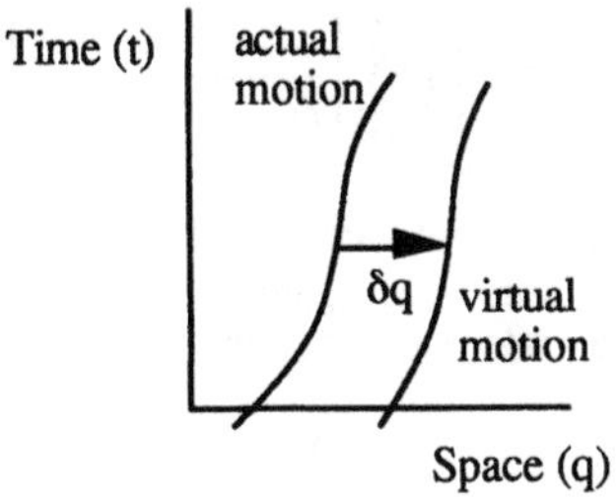


Fig. 4.01. Paths in extended configuration space

In Chapter II we introduced the idea of a virtual displacement in which we imagined subjecting, at fixed time t, each of the coordinates to a small change $\delta q_a$. At that stage in our discussion of mechanics, virtual displacements at time t and at a neighboring time $t + dt$ bore no particular relationship to one another. They were completely independent. We now restrict our attention to virtual displacements $\delta q_a$ which are continuous functions of time with continuous first and second time derivatives.[1] Surrounding any actual path corresponding to a possible motion of the system, we then have a family of neighboring virtual paths (which may or may not also be actual paths). We are interested on how certain physical quantities, functions of the generalized coordinates $q_a$ and generalized velocities $\dot{q}_a$, change as we go from a point on the actual path to a corresponding point on a neighboring virtual path. The change in the generalized coordinate $q_a$ is of course $\delta q_a$, but what is the change in the generalized velocity? To find this, consider

$$\frac{d}{dt}\delta q_a = \frac{d}{dt}(q_a + \delta q_a) - \frac{dq_a}{dt},$$

where we have added and subtracted $dq_a/dt$. Now the right-hand side is clearly the difference between the generalized velocity on the virtual path and the generalized velocity on the actual path. It is the change in velocity in going from the actual path to the corresponding point on the virtual path, thus

$$\frac{d}{dt}\delta q_a = \delta\frac{dq_a}{dt} = \delta\dot{q}_a.$$

This shows that the operations of actual displacement d and virtual displacement $\delta$ can be performed in either order; they "commute," $d\delta = \delta d$.

Let us now consider how the Lagrangian changes under a virtual displacement, as we go from the actual path to the corresponding point on a neighboring virtual path. We assume that L is a continuous function of the q's, $\dot{q}$'s, and t, with continuous first and second partial derivatives with respect to its arguments. We have

$$\delta L = \sum_{a=1}^{f}\left[\frac{\partial L}{\partial q_a}\delta q_a + \frac{\partial L}{\partial \dot{q}_a}\delta\dot{q}_a\right] = \sum_{a=1}^{f}\left[\frac{\partial L}{\partial q_a} - \frac{d}{dt}\left(\frac{\partial L}{\partial \dot{q}_a}\right)\right]\delta q_a + \frac{d}{dt}\sum_{a=1}^{f}\frac{\partial L}{\partial \dot{q}_a}\delta q_a.$$

For an actual path the generalized coordinates $q_a(t)$ satisfy Lagrange's equations. Hence the first term on the right-hand side is zero, and we are left with the second term, a total time derivative. If we now integrate both sides with respect to time from some initial time $t_0$ to some final time $t_1$, we get

$$\int_{t_0}^{t_1}\delta L\,dt = \int_{t_0}^{t_1}\frac{d}{dt}\sum_{a=1}^{f}\frac{\partial L}{\partial \dot{q}_a}\delta q_a\,dt = \sum_{a=1}^{f}\frac{\partial L}{\partial \dot{q}_a}\delta q_a\Bigg|_{t_0}^{t_1}.$$

The left-hand side can be written $\delta S$, where

[1] Less restrictive assumptions are sometimes possible; see, for example: R. Courant and D. Hilbert, *Methods of Mathematical Physics* (Interscience Publishers, Inc., New York, 1953), Vol. I, Chap. IV, in particular p. 200; I. M. Gelfand and S. V. Fomin, *Calculus of Variations*, (Prentice-Hall, Englewood Cliffs, NJ, 1963), trans. and ed. Richard A. Silverman, Sects. 4.1 and 15.

$$S[q] = \int_{t_0}^{t_1} L(q,\dot{q},t)\,dt,$$

the time integral of the Lagrangian along a path $q(t)$, is called the **action** for the path. Action is a type of quantity known as a **functional**, a quantity whose value is determined by specifying a function over a particular range. We shall see that the action functional $S[q]$ plays a major role in our subsequent studies in mechanics. The right-hand side of the preceding equation can be expressed in terms of the generalized momentum $p_a = \partial L/\partial \dot{q}_a$, so we finally obtain

$$\delta S = \sum_{a=1}^{f} p_a \delta q_a \Big|_{t_0}^{t_1}.$$

Now if the actual and virtual paths coincide at the initial and final times, so the virtual displacement $\delta q_a$ is zero at $t_0$ and $t_1$, we have

$$\delta S = 0.$$

The action is thus the *same* along the actual path and along any neighboring virtual path with the same end points. We say that the action along the actual path is **stationary**. Compare the situation in ordinary calculus: a function $f(x)$ is stationary (has a minimum, maximum, or point of inflection) at a particular point x if its change $df = (df/dx)dx$ in moving to a neighboring point $x + dx$ is zero.

We may now state:

**The principle of stationary action (Hamilton's principle)**

The actual path $q(t)$ of a mechanical system
between end points $(q_0, t_0)$ and $(q_1, t_1)$
is such that the action

$$S[q] = \int_{t_0}^{t_1} L(q,\dot{q},t)\,dt$$

is stationary
when compared with
neighboring virtual paths with the same end points.

We have derived the principle of stationary action from the equations of motion, from Lagrange's equations. But now we can turn the whole procedure around. We raise the principle of stationary action to the level of a *fundamental postulate* about the nature of mechanical motion. From it, by reversing the previous steps, we find

$$\delta S = \int_{t_0}^{t_1} \delta L\,dt = \int_{t_0}^{t_1} \sum_{a=1}^{f} \left[ \frac{\partial L}{\partial q_a} - \frac{d}{dt}\left( \frac{\partial L}{\partial \dot{q}_a} \right) \right] \delta q_a\,dt = 0.$$

Since at each instant the $\delta q_a$ represent arbitrary independent variations, the coefficients of each of them, the quantities in square brackets, must at each instant vanish.[2] The resulting equations are Lagrange's equations, which as we have seen are the equations of motion for a mechanical system in a wide range of circumstances.

The two statements about motion, "Lagrange's equations" and "principle of stationary action," are equivalent. However, the views they offer are quite different. "Lagrange's equations" are differential equations; they tell the system how to move one infinitesimal step at a time. "The principle of stationary action," on the other hand, is an integral principle; it requires the system to consider overall motions from start to finish, and to choose the one which makes the action integral stationary.

## Calculus of variations

The principle of stationary action is an example of a problem in the branch of mathematics known as the **calculus of variations**.[3] One is given a definite integral

$$I = \int \cdots \int F(y_i; \partial y_i/\partial x_\mu; \partial^2 y_i/\partial x_\mu \partial x_\nu; \cdots; x_\mu)\, dx_1 \cdots dx_N$$

whose integrand is a known function F of a set of dependent variables $y_i$, their derivatives to some order with respect to a set of independent variables $x_\mu$, and the independent variables $x_\mu$. The problem is to find the set of functions $y_i(x_1 \cdots x_N)$ which makes the integral stationary. If there is only one independent variable x, and if F does not contain derivatives of the dependent variables higher than first order, the required functions satisfy Lagrange's equations, usually known in this context as the **Euler-Lagrange equations**

$$\frac{d}{dx}\left(\frac{\partial F}{\partial(dy_i/dx)}\right) = \frac{\partial F}{\partial y_i}.$$

One of the earliest problems of this type to be considered was the so-called **brachistochrone problem**: find the curve joining two points in a uniform gravitational field such that the *time* required for a particle, starting from rest, to slide along the curve from the upper point to the lower is a minimum. The curve of minimum *distance* is, of course, the straight line connecting the points, but this is not necessarily the curve of minimum time. For example, if the start of the curve is steeper than the straight line, the particle will accelerate to a given speed more quickly, and this may more than compensate for the greater distance traveled. To set the problem up, choose the x-axis horizontal and the y-axis vertically down parallel to the gravitational field g. Suppose that the particle starts at the origin. Its speed when it has fallen a distance y is

$$v = \sqrt{2gy},$$

---

[2]This is not as obvious as it might at first appear, since the $\delta q$ are not really arbitrary but are restricted by certain continuity and smoothness conditions. For a proof see the references in footnote 1.

[3]Aspects of the theory beyond those required here can be found in the references of footnote 1.

and the time it takes to travel along the curve from $(0,0)$ to $(x_1,y_1)$ is given by

$$t = \int_0^{s_1} \frac{ds}{v},$$

where

$$ds = \sqrt{(dx)^2 + (dy)^2} = \sqrt{1 + (dy/dx)^2}\, dx$$

is the element of distance along the curve. The time can thus be written

$$t = \int_0^{x_1} \sqrt{\frac{1 + (dy/dx)^2}{2gy}}\, dx.$$

This has the same form as the expression for "action," with y playing the role of "generalized coordinate," x the role of "time," and

$$F(y, dy/dx) = \sqrt{\frac{1 + (dy/dx)^2}{2gy}}$$

the role of "Lagrangian." The curve $y(x)$ for minimum time satisfies the Euler-Lagrange equation. This equation can be integrated most readily by making use of a result proved in the next chapter (you may wish to prove the result before looking it up): if F does not depend explicitly on x, and if y satisfies the Euler-Lagrange equation, the quantity

$$H = \frac{dy}{dx}\frac{\partial F}{\partial(dy/dx)} - F \quad \left( = \frac{-1}{\sqrt{2gy(1 + (dy/dx)^2)}} \quad \text{in our case} \right)$$

is constant. We set

$$y(1 + (dy/dx)^2) = 2a$$

where a is a constant. This can be rearranged to give

$$\int_0^y \sqrt{\frac{y}{2a - y}}\, dy = x.$$

The integration is performed by setting

$$y = a(1 - \cos\phi).$$

We find

$$x = a\int_0^\phi (1-\cos\phi)\,d\phi = a(\phi - \sin\phi).$$

These last two equations give the required curve for minimum time, the "brachistochrone," in parametric form with x and y expressed in terms of a parameter $\phi$. The constant a must be chosen so that the curve passes through the end point $(x_1, y_1)$. See Fig. 4.02.

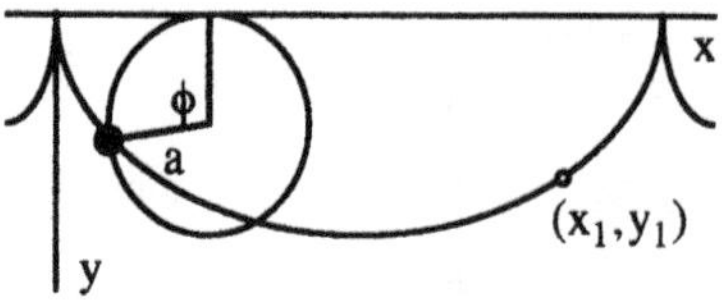


Fig. 4.02. The brachistochrone

The curve is a cycloid, which is the curve traced by a point on the rim of a rolling wheel. In our case the wheel has radius a and rolls along the x-axis. The starting point is a cusp of the cycloid. The parameter $\phi$ is the angle turned through. The center of the wheel is at $(a\phi, a)$, and the displacement of the point relative to the center is $(-a\sin\phi, -a\cos\phi)$.

## Geodesics

In Euclidean geometry a straight line is defined as the line of shortest distance between two points. A **geodesic** is the generalization for non-Euclidean curved spaces of a "straight line." In this section we derive the general equations for a geodesic in a curved space of an arbitrary number of dimensions. The results are useful not only in curved spaces, but also in Euclidean flat space when we choose to use curvilinear coordinates rather than cartesian coordinates. To find the equations, we must first learn how to write down an expression for the distance along a line. The distance ds between two neighboring points in Euclidean three-space, in terms of the cartesian coordinates $(x, y, z)$, is given by

$$(ds)^2 = (dx)^2 + (dy)^2 + (dz)^2.$$

As we have seen, it is frequently convenient to introduce, instead of cartesian coordinates, other coordinates

$$x^\alpha = x^\alpha(x, y, z)$$

which may be more suited to the physical problem at hand. The expression for the distance between two neighboring points in terms of these new coordinates can be obtained by changing variables. Since the expression for $(ds)^2$ is homogeneous quadratic in the cartesian coordinate differentials, it is homogeneous quadratic in the new coordinate differentials $dx^\alpha$ and is thus of the form

$$(ds)^2 = g_{\alpha\beta}\, dx^\alpha dx^\beta.$$

The set of quantities $g_{\alpha\beta}(x) = g_{\beta\alpha}(x)$, which in general are functions of the coordinates, form the components of the **metric tensor**. Here we have adopted, as is usual in this work, the Einstein **summation convention** whereby it is understood that repeated indices such as $\alpha$ and $\beta$ are to be summed over all values, whereas non-repeated or "free" indices can take on all values. In spherical polar coordinates $(r,\theta,\phi)$, for example, the distance between two neighboring points is given by

$$(ds)^2 = (dr)^2 + r^2(d\theta)^2 + r^2 \sin^2\theta\,(d\phi)^2,$$

so the components of the metric tensor are $g_{rr} = 1$, $g_{\theta\theta} = r^2$, $g_{\phi\phi} = r^2\sin^2\theta$, with all other components zero. While we have inferred the general form for the element of distance from that valid in Euclidean three-space, the expression holds for many other spaces, such as two-dimensional curved surfaces, (some) three-dimensional non-Euclidean spaces, and the four-dimensional space-time of relativity. Such spaces are called **Riemannian**.

A line is described by giving the coordinates $x^\alpha$ as functions

$$x^\alpha = x^\alpha(\lambda)$$

of some parameter $\lambda$. The distance between two neighboring points on the line is then

$$ds = \sqrt{g_{\alpha\beta}\frac{dx^\alpha}{d\lambda}\frac{dx^\beta}{d\lambda}}\, d\lambda,$$

and the distance along the line between two points with parameters $\lambda_0$ and $\lambda_1$ is given by

$$s = \int_{\lambda_0}^{\lambda_1} \sqrt{g_{\alpha\beta}\frac{dx^\alpha}{d\lambda}\frac{dx^\beta}{d\lambda}}\, d\lambda.$$

This is the expression for which we were looking. We now define a geodesic as a line joining two points, with the property that the distance along it is stationary when compared with neighboring lines which join the same points. The mathematical problem is the same as in the principle of stationary action. Taking over those results, we see that a geodesic is determined by the Euler-Lagrange equations with $\sqrt{g_{\alpha\beta}\frac{dx^\alpha}{d\lambda}\frac{dx^\beta}{d\lambda}}$ playing the role of "Lagrangian" and $\lambda$ playing the role of "time," thus

$$\frac{d}{d\lambda}\left(\frac{1}{\sqrt{g_{\alpha\beta}\frac{dx^\alpha}{d\lambda}\frac{dx^\beta}{d\lambda}}}g_{\mu\rho}\frac{dx^\rho}{d\lambda}\right)=\frac{1}{2}\frac{1}{\sqrt{g_{\alpha\beta}\frac{dx^\alpha}{d\lambda}\frac{dx^\beta}{d\lambda}}}\frac{\partial g_{\rho\sigma}}{\partial x^\mu}\frac{dx^\rho}{d\lambda}\frac{dx^\sigma}{d\lambda}.$$

Recall that the repeated indices $\alpha,\beta,\rho,\sigma$ are summed over, and the free index $\mu$ can take on all possible values. Let us now choose the parameter $\lambda$ to be the arc-length s along the line. We then have the constraint

$$g_{\alpha\beta}\frac{dx^\alpha}{ds}\frac{dx^\beta}{ds}=1,$$

and the equations simplify to

$$\frac{d}{ds}\left(g_{\mu\rho}\frac{dx^\rho}{ds}\right)=\frac{1}{2}\frac{\partial g_{\rho\sigma}}{\partial x^\mu}\frac{dx^\rho}{ds}\frac{dx^\sigma}{ds}.$$

Carrying out the differentiation on the left and rearranging, we find

$$g_{\mu\rho}\frac{d^2x^\rho}{ds^2}=-\frac{1}{2}\left(2\frac{\partial g_{\mu\rho}}{\partial x^\sigma}-\frac{\partial g_{\rho\sigma}}{\partial x^\mu}\right)\frac{dx^\rho}{ds}\frac{dx^\sigma}{ds}=-\frac{1}{2}\left(\frac{\partial g_{\mu\rho}}{\partial x^\sigma}+\frac{\partial g_{\mu\sigma}}{\partial x^\rho}-\frac{\partial g_{\rho\sigma}}{\partial x^\mu}\right)\frac{dx^\rho}{ds}\frac{dx^\sigma}{ds}.$$

The second equality follows from noting that $\frac{dx^\rho}{ds}\frac{dx^\sigma}{ds}$ is symmetric in $\rho$ and $\sigma$, and hence the term which multiplies it can be taken symmetric as well. The resulting equations for a geodesic are usually written

$$g_{\mu\rho}\frac{d^2x^\rho}{ds^2}=-\Gamma_{\mu,\rho\sigma}\frac{dx^\rho}{ds}\frac{dx^\sigma}{ds}$$

where

$$\Gamma_{\mu,\rho\sigma}=\frac{1}{2}\left(\frac{\partial g_{\mu\rho}}{\partial x^\sigma}+\frac{\partial g_{\mu\sigma}}{\partial x^\rho}-\frac{\partial g_{\rho\sigma}}{\partial x^\mu}\right)$$

are the components of the **Christoffel symbol**. These equations for a geodesic by themselves imply that $g_{\alpha\beta}\frac{dx^\alpha}{ds}\frac{dx^\beta}{ds}$ is constant, and the constraint requires that this constant be taken to be 1.

**Examples:**

1. In Euclidean three-space and with cartesian coordinates, the components of the Christoffel symbol are zero, and the coordinates satisfy

$$\frac{d^2x^\mu}{ds^2} = 0.$$

They are thus linear functions of s,

$$\mathbf{r} = \mathbf{r}_0 + \mathbf{n}s$$

with $|\mathbf{n}| = 1$. These, of course, are the equations of a straight line; the components of **n** are the direction cosines of the line.

2. On the (two-dimensional) surface of a sphere of radius R and with spherical polar coordinates $\theta$ and $\phi$ (the co-latitude and longitude), the non-vanishing components of the Christoffel symbol are

$$\Gamma_{\theta,\phi\phi} = -\Gamma_{\phi,\theta\phi} = -\Gamma_{\phi,\phi\theta} = -R^2 \sin\theta\cos\theta,$$

and the equations for a geodesic become

$$\frac{d^2\theta}{ds^2} = \sin\theta\cos\theta\left(\frac{d\phi}{ds}\right)^2 \qquad \frac{d^2\phi}{ds^2} = -2\cot\theta\frac{d\theta}{ds}\frac{d\phi}{ds}$$

$$\text{with} \quad R^2\left[\left(\frac{d\theta}{ds}\right)^2 + \sin^2\theta\left(\frac{d\phi}{ds}\right)^2\right] = 1.$$

To find the solutions to these equations, we first note that if $d\phi/ds$ starts off zero, it remains zero and the equations reduce to

$$\frac{d\phi}{ds} = 0 \qquad \frac{d\theta}{ds} = \pm\frac{1}{R}$$

with solutions

$$\phi = \phi_0 \qquad \theta = \theta_0 \pm (s/R).$$

These are lines of longitude and are suitable geodesics when the end points lie along a north-south line. If this is not the case, it is more convenient to describe the geodesics by giving $\theta$ as a function of $\phi$. To do this, we set

$$\frac{d\theta}{ds} = \frac{d\theta}{d\phi}\frac{d\phi}{ds} \qquad \frac{d^2\theta}{ds^2} = \frac{d^2\theta}{d\phi^2}\left(\frac{d\phi}{ds}\right)^2 + \frac{d\theta}{d\phi}\frac{d^2\phi}{ds^2}$$

in the equations for a geodesic and hence find

$$\frac{d^2\theta}{d\phi^2} - 2\cot\theta\frac{d\theta}{d\phi} = \sin\theta\cos\theta.$$

This can be written

$$\frac{d^2}{d\phi^2}\cot\theta = -\cot\theta,$$

which has the form of a "simple harmonic oscillator equation" in $\cot\theta$, with solution $\cot\theta \propto \cos(\phi - \phi_0)$. It is convenient to choose the constant of proportionality to be $-\tan\theta_0$. The solution, the equation for a geodesic on a sphere, then becomes

$$\sin\theta\sin\theta_0\cos(\phi - \phi_0) + \cos\theta\cos\theta_0 = 0.$$

If multiplied by the radial distance r, this is the equation of a plane through the center of the sphere; note that the equation can be written $\mathbf{n}\cdot\mathbf{r} = 0$, where $\mathbf{r} = (r\sin\theta\cos\phi, r\sin\theta\sin\phi, r\cos\theta)$ gives the cartesian coordinates of a point on the plane, and $\mathbf{n} = (\sin\theta_0\cos\phi_0, \sin\theta_0\sin\phi_0, \cos\theta_0)$ gives the cartesian components of the unit normal to the plane. The geodesics are the intersections of this family of planes with the surface $r = R$ of the sphere. They are the **great circle** routes. Two such routes connect any two (non-antipodal) points on the sphere. The shorter is the shortest route between the two points, but the longer is neither a route of (local) minimum distance or of maximum distance: there are nearby routes which, when second order terms are included, are shorter than the long great circle, and also routes which are longer than the long great circle.

This is a good point to comment on the circumstances under which paths of *stationary* action are paths of *least* action. Consider a path C of stationary action from point 0 to point 1. This path leaves 0 with some particular velocity. Paths which leave 0 with slightly different initial velocity initially spread out. However, some or all may eventually come together again, intersecting the path C at what are known as points conjugate to 0. It can be shown that C is a path of *least* action provided 1 is closer to 0 than the first conjugate point of 0.[4]

3. In the special and general theories of relativity, **events** in space-time are labeled by four coordinates $x^\mu$, with $x^0 = ct$ the time and $(x^1, x^2, x^3)$ the three space coordinates. The invariant **interval** between two neighboring events is given by

$$(ds)^2 = g_{\alpha\beta}\, dx^\alpha dx^\beta,$$

where the metric tensor, which determines the geometry of space-time, is itself determined by the energy-momentum distribution of matter (including fields) in the space.

---

[4] I. M. Gelfand and S. V. Fomin, *Calculus of Variations*, (Prentice-Hall, Englewood Cliffs, NJ, 1963), trans. and ed. Richard A. Silverman, Chap. 5.

In the flat space-time of special relativity and with cartesian coordinates, the interval becomes

$$(ds)^2 = c^2(dt)^2 - (dx)^2 - (dy)^2 - (dz)^2.$$

The geodesics, which are the paths of a free particle, are straight lines

$$t = t_0 + \gamma\tau$$
$$\mathbf{r} = \mathbf{r}_0 + \gamma\mathbf{v}\tau.$$

Here $\tau = s/c$ is the proper time as measured by a clock carried by the particle, $\mathbf{v} = d\mathbf{r}/dt$ is the velocity of the particle, and $\gamma^{-1} = \sqrt{1-(v/c)^2}$ is the **Lorentz contraction factor**. Imagine the particle to be a rocket ship going from the earth to the moon, and take the x-axis in the earth-moon direction. The earth-moon displacement $v\tau$ as determined by rocket observers is a factor $\gamma^{-1}$ less than the earth-moon displacement $x - x_0$ as determined by earth-moon observers. Rocket observers can picture the earth-moon displacement as a measuring rod which moves with velocity $-v$ with respect of them, and they find the length of this moving rod to be *less* than that found in the rod's rest, earth-moon, frame; the moving rod is short. This is known as **Lorentz contraction**. Also, the time interval $\tau$ between the two events, rocket leaves earth and rocket arrives at moon, as determined by rocket observers is a factor $\gamma^{-1}$ less than the time interval $t - t_0$ as determined by earth-moon observers. Alternatively, earth-moon observers find the time interval between the two events to be a factor $\gamma$ *greater* than the time interval found by the moving, rocket, clock which is coincident with the two events; the moving clock runs slow. This is known as **time dilation**.

In general relativity, motions, which the Newtonian view would attribute to the effects of "gravity," are attributed to "space-time curvature."[5] In particular, particles which Newton would say are "acted on only by gravity" follow geodesic lines through curved space-time. In order to see how these views connect, let us consider a limiting case in which the gravitational field is weak, so the components of the metric tensor deviate little from their special relativity values, and let us further assume that the velocity of the particles is small compared to that of light. We then have $(dx^0/ds) = c(dt/ds) \approx 1$ and $(d\mathbf{r}/ds) \approx (v/c) << 1$ and can thus ignore on the right-hand side of the equation for a geodesic all terms in the sum over $\rho$ and $\sigma$ except the one for which $\rho = \sigma = 0$. The equation for the x-component becomes

$$-\frac{1}{c^2}\frac{d^2x}{dt^2} \approx -\Gamma_{1,00} = +\frac{1}{2}\frac{\partial g_{00}}{\partial x}, \quad \text{which generalizes to} \quad \frac{d^2\mathbf{r}}{dt^2} \approx -\nabla\left(\tfrac{1}{2}c^2 g_{00}\right).$$

[5] A. Einstein in H. A. Lorentz, A. Einstein, H. Minkowski, and H. Weyl, *The Principle of Relativity* (Dover Publications, New York, NY, 1923), trans. W. Perrett and G. B. Jeffery, p. 111; Wolfgang Rindler, *Essential Relativity*, (Springer-Verlag, New York, NY, 1969; 1977;1979), rev. 2nd. ed.; Ronald Adler, Maurice Bazin, and Menachem Schiffer, *Introduction to General Relativity*, (McGraw-Hill Book Company, New York, NY, 1965, 1975), 2nd ed.; Steven Weinberg, *Gravitation and Cosmology: Principles and Applications of the General Theory of Relativity*, (John Wiley and Sons, New York, NY, 1972).

This is Newton's equation of motion for a particle moving in a gravitational potential (gravitational potential energy per unit mass)

$$\phi = \text{constant} + \tfrac{1}{2}c^2 g_{00}.$$

We choose the constant so that $g_{00}$ reduces to its flat space-time value +1 far from matter, where we take, as usual, $\phi = 0$. The preceding equation then gives the time-time component of the metric tensor

$$g_{00}(\mathbf{r}) = 1 + \frac{2\phi(\mathbf{r})}{c^2}$$

as modified by a weak gravitational field. Outside a spherical distribution of mass M the gravitational potential is $\phi = -GM/r$, so, for example, the correction at the surface of the earth is $-1.4 \times 10^{-9}$.

One of the consequences of this modification to the metric tensor is that clocks in a gravitational field run slow. To see this, first note that (for a static gravitational field and suitable coordinates) if a light pulse leaves point A at coordinate time $t_A$ and arrives at point B at time $t_B$, then a light pulse which leaves point A at time $t'_A$ will arrive at point B at time $t'_B$ such that $t'_B - t'_A = t_B - t_A$. The "travel time delay" remains constant. This can be rearranged to give $t'_B - t_B = t'_A - t_A$, which says that the time interval between the pulses as indicated on the coordinate clocks at A and B is the same. The rates of the coordinate clocks have been adjusted so that this is so. On the other hand, the time interval $\Delta\tau$ $(= \Delta s/c)$ indicated by a stationary standard clock at $\mathbf{x}$, such as a particular atom emitting a particular spectral line, is related to the coordinate time interval $\Delta t$ by $\Delta\tau = \sqrt{g_{00}(\mathbf{x})}\,\Delta t$. The interval between the two pulses as indicated by a standard clock at A (proportional to the number of periods of the given spectral line) is thus $\Delta\tau_A = \sqrt{g_{00}(A)}\,\Delta t_A$, and that indicated by a standard clock of identical construction at B is $\Delta\tau_B = \sqrt{g_{00}(B)}\,\Delta t_B$. Since $\Delta t_B = \Delta t_A$, we have

$$\frac{\Delta\tau_B}{\Delta\tau_A} = \sqrt{\frac{g_{00}(B)}{g_{00}(A)}} \approx 1 + \frac{\phi_B - \phi_A}{c^2},$$

the second equality following from our expression for $g_{00}$. If B is at a higher gravitational potential than A, then $\Delta\tau_B > \Delta\tau_A$. It takes more periods for the standard clock at B to fill the interval between the pulses than for the standard clock at A; the standard clock at A thus runs slow compared to that at B. Another way to say this is in terms of frequency. Let $\Delta\tau_A$ be the period and $\nu_{AA} = 1/\Delta\tau_A$ the frequency of a particular spectral line of an atom at A, as determined at A. If this light travels to B, its period as determined there is $\Delta\tau_B$ as given above, and its frequency there is $\nu_{AB} = 1/\Delta\tau_B$. The frequency of the same spectral line of an identical atom at B, as determined at B, is of course $\nu_{BB} = \nu_{AA}$. We thus have

$$\frac{\nu_{AB}}{\nu_{BB}} = \sqrt{\frac{g_{00}(A)}{g_{00}(B)}} \approx 1 + \frac{\phi_A - \phi_B}{c^2}.$$

If B is at a higher gravitational potential than A, then $\nu_{AB} < \nu_{BB}$ and light from A is red-shifted compared to that from B.[6]

## Path integral formulation of quantum mechanics[7]

Consider the following experiment: a beam of electrons is emitted by a source, passes through a double slit, and is observed on a screen (Fig. 4.03).

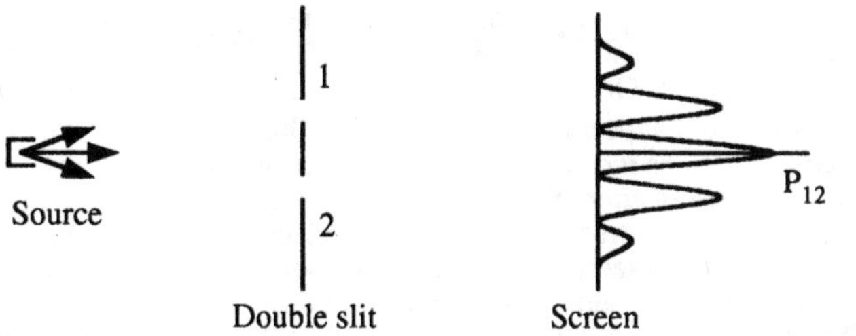


Fig. 4.03. The double slit experiment

The results are:

1. The electrons arrive at any given spot on the screen in bits of fixed mass and charge at some spot-dependent average rate; if we decrease the intensity of the beam, the rate decreases, but the size of the bits stays the same. We may say: "electrons behave like particles."

2. We can measure the probability for an electron to arrive at various spots on the screen. The result is curve $P_{12}$. This looks just like the double slit interference pattern for classical waves, and based on this we may say: "electrons behave like waves."

Our aim is to try to understand these two apparently contradictory results. To do so, we must be extraordinarily careful in what we say and in the mental pictures we create. For example, if "electrons behave like particles," we might be tempted to say: "electrons which arrive at the screen must have passed *either* through slit 1 *or* through slit 2." To check this, we can block off slit 2 and measure the probability of arrival for electrons which have passed through slit 1. The result is curve $P_1$ (Fig. 4.04). Repeat for slit 1 blocked off, and the result is curve $P_2$.

---

[6]R. V. Pound and G. A. Rebka, "Apparent Weight of Photons," Phys. Rev. Lett. **4**, 337-341 (1960); R. V. Pound and J. L. Snider, "Effect of Gravity on Nuclear Resonance," Phys. Rev. Lett. **13**, 539-540 (1964).

[7]For parallel reading see: R. P. Feynman, "Space-Time Approach to Non-Relativistic Quantum Mechanics," Rev. Mod. Phys. **20**, 367-387 (1948); Richard P. Feynman, Robert B. Leighton, Matthew Sands, *The Feynman Lectures on Physics*, vol. 3, (Addison-Wesley Publishing Company, Reading, MA, 1965); R. P. Feynman and A. R. Hibbs, *Quantum Mechanics and Path Integrals*, (McGraw-Hill Book Company, New York, NY, 1965).

Fig. 4.04. Probability with one slit blocked

It is clear that $P_{12} \neq P_1 + P_2$. Note in particular that for $P_{12}$ there are points where no electrons arrive when both slits are open, so it appears that closing one slit has *increased* the number from the other. On the other hand, note that there are points (say the center) where $P_{12} > P_1 + P_2$, so it appears that closing one slit has *decreased* the number from the other. These considerations indicate that it is improper to say: "electrons which arrive at the screen must have passed *either* through slit 1 *or* through slit 2."

But suppose we keep track of which slit the electrons go through, and where they end up on the screen. For example, we could put a source of light between the slits and observe the scattered light. It turns out that we observe scattered light *either* near slit 1 *or* near slit 2. It would appear that the electrons indeed go through one slit or the other; they behave like particles. Further, electrons which go through 1 give curve $P_1$, electrons which go through 2 give curve $P_2$, and electrons which go through 1 or 2 necessarily give curve $P_1 + P_2$. There is now no interference! The process of observing which slit the electron goes through destroys the interference pattern. Suppose we try to minimize the effect of the observing by turning down the intensity of the light.[8] Then, however, we sometimes miss an electron; it gets to the screen without being observed, so we can't tell which slit it came through. And these electrons give curve $P_{12}$!

We cannot form a classical picture of this, but we can describe mathematically what happens as follows:

1. There is a certain **probability amplitude** $\phi_1$ (a complex number) for arriving if slit 1 is open, such that $P_1 = |\phi_1|^2$. Ditto for 2.

2. The probability amplitude for arriving with slits 1 and 2 open, and no detection apparatus present, is $\phi_{12} = \phi_1 + \phi_2$, and

$$\begin{aligned} P_{12} &= |\phi_1 + \phi_2|^2 \\ &= |\phi_1|^2 + |\phi_2|^2 + 2\,\mathrm{Re}(\phi_1^* \phi_2) \\ &= P_1 + P_2 + 2\sqrt{P_1 P_2}\cos\delta \end{aligned}$$

where $\delta$ is the phase difference between the two amplitudes. Because of the last term, there is probability interference. If there is some apparatus present which is capable of

[8]We may also decrease the frequency of the light. What happens then?

determining which slit the electron goes through, its effect is to randomize the phase difference $\delta$. The last term then gives zero, and we recover the classical result.

The double slit experiment suggests the following generalization: the probability for an electron to go from one space-time point A to another space-time point B is the square of the modulus of a probability amplitude $\phi$. This amplitude can be written as a sum over all the various paths leading from A to B of the probability amplitude $\phi$[path] associated with each path (Fig. 4.05).

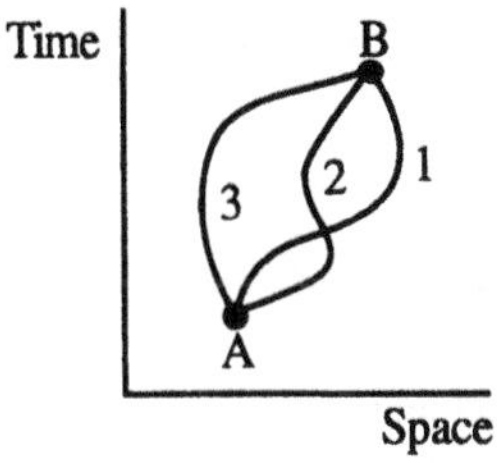


Fig. 4.05. Some paths between A and B

$$\phi = \phi_1 + \phi_2 + \phi_3 + \cdots = \sum_{\text{paths}} \phi[\text{path}]\ .$$

According to Feynman the amplitude for a particular path is given by

$$\phi[\text{path}] \propto e^{\frac{i}{\hbar}S[\text{path}]}\ ,$$

where S[path] is the classical action for the path and $\hbar$ is Planck's constant (divided by $2\pi$). The proportionality constant, not written in the above equation, is the same for all paths. The sum over paths is called a **Feynman path integral** and the approach based on it is known as the path integral formulation of quantum mechanics. Let us see how it works.

First consider the classical limit: the results of quantum mechanics should approach those of classical mechanics in the limit in which the action S is large compared to $\hbar$. In this limit the factor $\exp\left(\frac{i}{\hbar}S[\text{path}]\right)$ depends very sensitively on S. It oscillates extremely rapidly with small changes in S, so contributions to the amplitude from neighboring paths with slightly different S tend to cancel. The only paths which contribute significantly are those for which neighboring paths have the *same* action S, that is those for which

$$\delta S = 0\ .$$

As we have seen, this is Hamilton's principle, which gives the classical path.

Next consider how the path integral formulation connects with ordinary quantum mechanics, and in particular with the **Schrödinger equation**. This is an equation for the **wave function** $\psi(x,t)$, the amplitude for finding the particle at x at time t. (The probability for finding the particle in the interval x to $x + dx$ at time t is $|\psi(x,t)|^2\,dx$.) The time-dependent Schrödinger equation relates the amplitude at time t to the amplitude at a slightly earlier time; it gives the time derivative of the amplitude. In the path integral formulation the amplitude $\psi(x,t)$ for finding the particle at x at time t equals the amplitude $\psi(x-\xi,t-\varepsilon)$ for finding it at a point $x-\xi$ at a slightly earlier time $t-\varepsilon$, multiplied by the amplitude for it to go from $(x-\xi,t-\varepsilon)$ to $(x,t)$, summed over all initial positions. We are interested in the limit $\varepsilon \to 0$. In this limit the amplitude for the particle to go from $(x-\xi,t-\varepsilon)$ to $(x,t)$ can be approximated by the amplitude for it to travel along the single constant velocity path connecting the two points. We then have

$$\psi(x,t) \approx \int_{-\infty}^{\infty} \frac{d\xi}{A} e^{\frac{i}{\hbar}S}\,\psi(x-\xi,t-\varepsilon),$$

where A is a normalization constant and

$$S = \left[\tfrac{1}{2}m(\xi/\varepsilon)^2 - V(x)\right]\varepsilon$$

is the action, to lowest order in $\varepsilon$, for the constant velocity path connecting the points $(x-\xi,t-\varepsilon)$ and $(x,t)$. We thus have

$$\psi(x,t) \approx \int_{-\infty}^{\infty} \frac{d\xi}{A} e^{\frac{im\xi^2}{2\hbar\varepsilon}} e^{-\frac{i}{\hbar}V(x)\varepsilon}\,\psi(x-\xi,t-\varepsilon).$$

In the limit $\varepsilon \to 0$ the factor $e^{\frac{im\xi^2}{2\hbar\varepsilon}}$ oscillates extremely rapidly, except when $|\xi|$ is small, less than something of the order of $\sqrt{\hbar\varepsilon/m}$. The dominant contribution to the $\xi$ integration comes from this small region about $\xi = 0$. We can thus expand the other factors in the integrand in a Taylor series in $\varepsilon$ and $\xi$, retaining terms up to first order in $\varepsilon$ and (to be consistent) to second order in $\xi$,

$$\psi(x,t) \approx \int_{-\infty}^{\infty} \frac{d\xi}{A} e^{\frac{im\xi^2}{2\hbar\varepsilon}} \left[\psi - \frac{\partial\psi}{\partial t}\varepsilon - \frac{i}{\hbar}V\psi\varepsilon + \cdots - \frac{\partial\psi}{\partial x}\xi + \frac{1}{2}\frac{\partial^2\psi}{\partial x^2}\xi^2 + \cdots\right],$$

where the wave function and its derivatives are now all evaluated at $(x,t)$. The integration over $\xi$ can be performed with the aid of the (Fresnel) integrals

$$\int_{-\infty}^{\infty} d\xi\, e^{ia\xi^2} = \sqrt{\frac{\pi i}{a}}, \quad \int_{-\infty}^{\infty} d\xi\,\xi\, e^{ia\xi^2} = 0, \quad \int_{-\infty}^{\infty} d\xi\,\xi^2 e^{ia\xi^2} = \frac{i}{2a}\sqrt{\frac{\pi i}{a}},$$

with the result

$$\psi(x,t) \approx \frac{1}{A}\sqrt{\frac{2\pi i\hbar\varepsilon}{m}}\left[\psi + \left(-\frac{\partial\psi}{\partial t} - \frac{i}{\hbar}V\psi + \frac{i\hbar}{2m}\frac{\partial^2\psi}{\partial x^2}\right)\varepsilon + \cdots\right].$$

To make the zeroth order terms in $\varepsilon$ agree, we must choose the normalization constant A to be

$$A = \sqrt{\frac{2\pi i\hbar\varepsilon}{m}}.$$

The first order terms in $\varepsilon$ then give

$$i\hbar\frac{\partial\psi}{\partial t} = -\frac{\hbar^2}{2m}\frac{\partial^2\psi}{\partial x^2} + V\psi,$$

which is the time-dependent Schrödinger equation.

## Exercises

1. Consider a modified brachistochrone problem in which the particle has non-zero initial speed $v_0$. Show that the brachistochrone is again a cycloid, but with cusp $h = v_0^2/2g$ higher than the initial point.

2. A bead of mass m slides without friction along a wire bent in the shape of a cycloid
$$x = a(\phi - \sin\phi) \qquad y = a(1 - \cos\phi).$$
Gravity g acts vertically down, parallel to the y axis.
(a) Find the displacement s along the cycloid, measured from the bottom, in terms of the parameter $\phi$.
(b) Write down the Lagrangian using s as generalized coordinate, and show that the motion is simple harmonic in s with period independent of amplitude. Thus the time required for the bead, starting from rest, to slide from any point on the cycloid to the bottom is independent of the starting point. What is this time?

3. Novelists have long been fascinated with the idea of a worldwide rapid transit system consisting of subterranean passages crisscrossing the earth.[9] Public interest in subterranean travel rose sharply when *Time* magazine[10] commented on a paper by Paul W. Cooper, "Through the Earth in Forty Minutes".[11] This paper, while

---

[9]See Martin Gardner, *Scientific American*, September 1965, pp. 10-12, commenting on an article by L. K. Edwards, "High-Speed Tube Transportation," *Scientific American*, August 1965, pp. 30-40.
[10]*Time*, February 11, 1966, pp. 42-43.
[11]Paul W. Cooper, "Through the Earth in Forty Minutes," Am. J. Phys. **34**, 68-70 (1966).

repeating some earlier work,[12] served as a catalyst for a number of other papers on the subject[13] to which you may wish to refer in working the present exercise. Take the gravitational potential within the earth to be $\frac{1}{2}(g/R)r^2$ where g is the gravitational field at the surface and R is the radius of the earth (thereby neglecting the non-uniform density of the earth).

(a) First show that a particle starting from rest and sliding without friction through a *straight* tunnel connecting two points on the surface of the earth executes simple harmonic motion, and that the time to slide from one end to the other is $\tau_0 = \pi\sqrt{R/g}$ ($\approx 42.2\,\text{min}$) independent of the location of the end points.

(b) Now consider the curve $r(\theta)$ the tunnel must follow such that the time for the particle to slide from one end to the other is minimum. Set up the appropriate variational principle, and show that

$$\frac{r^2}{\sqrt{(dr/d\theta)^2 + r^2}\sqrt{R^2 - r^2}} = \frac{r_0}{\sqrt{R^2 - r_0^2}}$$

is a first integral of the resulting Euler-Lagrange equation. Here $r = r_0$ at the bottom of the tunnel ($r_0$ is the minimum distance to the center of the earth). Rearrange this and integrate to obtain the equation of the curve,

$$\theta = \tan^{-1}\left(\frac{R}{r_0}\sqrt{\frac{r^2 - r_0^2}{R^2 - r^2}}\right) - \frac{r_0}{R}\tan^{-1}\left(\sqrt{\frac{r^2 - r_0^2}{R^2 - r^2}}\right),$$

where $\theta$ is measured from the bottom of the tunnel. The angular separation between the end points on the surface of the earth is thus given by

$$\Delta\theta = \pi(1 - r_0/R).$$

(c) Introduce a parameter $\phi$ with

$$\tan\frac{\phi}{2} = \sqrt{\frac{r^2 - r_0^2}{R^2 - r^2}},$$

so $\phi = 0$ at the bottom and $\phi = \pm\pi$ at the ends of the tunnel. Show that the equation of the curve takes the form

$$r^2 = \tfrac{1}{2}\left(R^2 + r_0^2\right) - \tfrac{1}{2}\left(R^2 - r_0^2\right)\cos\phi$$

$$\theta = \tan^{-1}\left(\frac{R}{r_0}\tan\frac{\phi}{2}\right) - \frac{r_0}{2R}\phi\,.$$

Show that this is the equation of a hypocycloid, which is the curve traced by a point on the circumference of a circle which rolls without slipping on another circle.

---

[12]See Philip G. Kirmser, "An Example of the Need for Adequate References," Am. J. Phys. **34**, 701 (1966).

[13]Giulio Venezian, "Terrestrial Brachistochrone," Am. J. Phys. **34**, 701 (1966); Russell L. Mallett, "Comments on 'Through the Earth in Forty Minutes'," Am. J. Phys. **34**, 702 (1966); L. Jackson Laslett, "Trajectory for Minimum Transit Time Through the Earth," Am. J. Phys. **34**, 702-703 (1966); Paul W. Cooper, "Further Commentary on 'Through the Earth in Forty Minutes'," Am. J. Phys. **34**, 703-704 (1966).

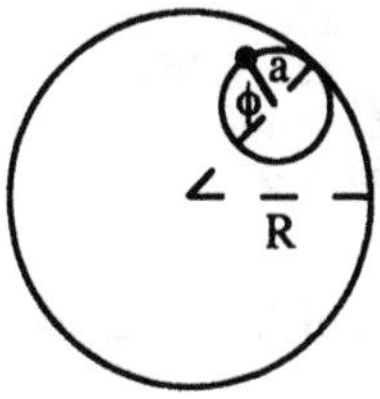


In this case the larger circle is the great circle route, of radius R, connecting the end points on the surface of the earth, and the smaller circle has radius $a = \frac{1}{2}(R - r_0)$ (its circumference is thus the distance between the end points on the surface). The parameter $\phi$ is the angle shown in the figure.

(d) Now consider the time dependence of the variables. Show in particular that $\phi$ varies linearly with time, $\phi = 2\pi(t/\tau)$, where $\tau = \tau_0\sqrt{1-(r_0/R)^2}$ is the time to slide through the minimum-time-tunnel from one end to the other. Compare $\tau$ with $\tau_0$ for end points 700 km apart on the surface.

4. An instructive exercise in the calculus of variations is the "minimum surface of revolution problem":

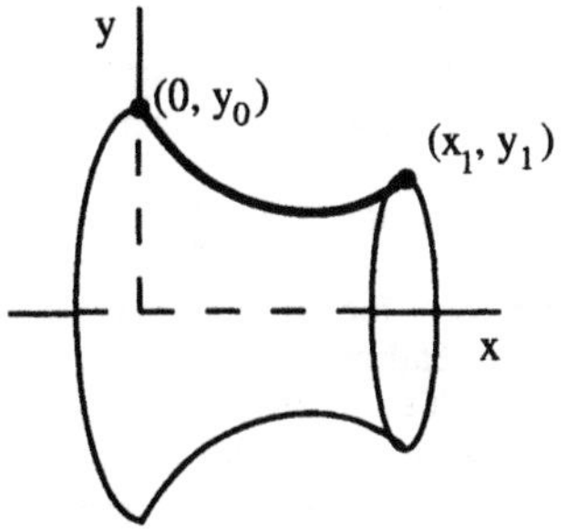


(a) Find the plane curve $y = y(x)$ joining two points $(0, y_0)$ and $(x_1, y_1)$ such that the area of the surface formed by rotating the curve about the x-axis is minimum (the Euler-Lagrange answer is $y = a\cosh((x - b)/a)$, where a and b must be chosen so that the curve passes through the end points).

(b) Using a computer or otherwise, draw representative members of the (one-parameter) family of such curves which start at $(0,1)$. Hence convince yourself that if the final point is near the y-axis, *two* Euler-Lagrange curves pass through the given end points, whereas if the final point is near the x-axis, *no* Euler-Lagrange curves pass through the given end points.

(c) In this latter case the solution is the discontinuous **Goldschmidt solution** composed of straight line segments $(0,1) \to (0,0) \to (x_1,0) \to (x_1,y_1)$. In the region where there are two Euler-Lagrange solutions, calculate and compare the area given by the Goldschmidt solution with the areas given by the Euler-Lagrange solutions. Which of the three gives minimum area? Does this depend on where in the region the end point lies? (This last part is difficult; for guidance see Gilbert Ames Bliss, *Calculus of Variations*, (published for The Mathematical Association of America by The Open Court Publishing Company, Chicago, Illinois, 1925), Chap. IV, pp. 85-127.)

5. The motion of a "free" particle of mass m on a surface is described by Lagrange's equations with Lagrangian $L = T = \frac{1}{2}m(ds/dt)^2$. Show that the resulting equations of motion are the equations for a geodesic, along which the particle moves at constant speed $ds/dt$.

6. Find and solve the equations for geodesics on a plane, using plane polar coordinates $(r,\phi)$ in terms of which the element ds of distance is given by $ds^2 = dr^2 + r^2 d\phi^2$.

7. (a) Find and solve the equations for geodesics on the surface of a cone of half angle $\alpha$, using as coordinates the distance r from the apex of the cone and the azimuthal angle $\phi$.
(b) Show that if the cone is cut along a line $\phi$ = constant and flattened out onto a plane, the geodesics become straight lines.

8. Consider two points on the surface of a sphere. Without loss we may take them on the equator at $(\theta = \pi/2, \phi = 0)$ and $(\theta = \pi/2, \phi = \alpha)$. The geodesics joining these points are the two arcs of the equator. Nearby curves can be represented by

$$\theta = \pi/2 + \sum_{n=1}^{\infty} a_n \sin(\pi n\phi/\alpha)$$

where the deviation from the equator has been represented by a Fourier series chosen to vanish at the end points. Evaluate the distance between the two points along such a curve, valid to second order in the small quantities a. Show that for $0 < \alpha < \pi$ the distance is always longer than the distance along the equator, whereas for $\pi < \alpha < 2\pi$ there are nearby curves for which the distance is shorter than that along the equator, as well as ones for which the distance is longer than that along the equator.

9. (a) Evaluate the action $S[\mathbf{x}(t)]$ for a free particle along the path:
"from $(\mathbf{x}_0, t_0)$ to $(\mathbf{x}', t')$ at constant velocity, and then
from $(\mathbf{x}', t')$ to $(\mathbf{x}_1, t_1)$ at (a usually different) constant velocity."
(b) Consider S as a function of a parameter $\mathbf{x}'$. Show that minimum action results when $\mathbf{x}'$ is chosen so that the velocity from $(\mathbf{x}_0, t_0)$ to $(\mathbf{x}', t')$ is the same as that from $(\mathbf{x}', t')$ to $(\mathbf{x}_1, t_1)$, so that the full motion is at constant velocity.

10. **Fermat's principle** states that light travels from one point to another along the trajectory which makes the travel time a minimum.
(a) Use Fermat's principle to derive the law for the reflection of light from a mirror, namely

"angle of incidence = angle of reflection"

(b) Use Fermat's principle to derive Snell's law for the refraction of light passing from a medium in which the speed of light is $c/n_0$ to a medium in which the speed of light is $c/n_1$ (c is the speed of light in free space and n is the index of refraction), namely

$$n_0 \sin\phi_0 = n_1 \sin\phi_1.$$

Here $\phi_0$ and $\phi_1$ are the angles to the normal of the incident and refracted rays.

11. **Jacobi's principle** states that a particle of mass m and energy E in a potential $V(x,y,z)$ travels from one point to another along a trajectory which makes the integral $\int p(x,y,z)\,ds$, where $p(x,y,z) = \sqrt{2m(E - V(x,y,z))}$ is the magnitude of the momentum, stationary (for further details, see Chapter VIII).

(a) Consider projectile motion in the $(x,y)$ plane with x horizontal and y vertical, and with potential $V = mgy$ where g is the (constant) gravitational field. Write down the Euler-Lagrange equation which results from Jacobi's principle, and integrate to obtain the equation for the trajectories.

(Ans. $y - y_0 = (x - x_0)\tan\alpha - \dfrac{(x - x_0)^2}{4h\cos^2\alpha}$ where $(x_0, y_0)$ is the start point, $E = mgh$ is the energy, and $\alpha$ is the angle of launch)

(b) Sketch the family of trajectories which start at $(0,0)$ with fixed energy $E = mgh$ but arbitrary angle $\alpha$ of launch. Show that if the end point lies within the envelope $y = h - (x^2/4h)$ of the family of trajectories, two trajectories connect the start and end points, whereas if the end point lies outside the envelope, no trajectories connect the start and end points.

12. A particle moves vertically in the uniform gravitational field g near the surface of the earth. The Lagrangian is

$$L = \tfrac{1}{2}m\dot{z}^2 - mgz.$$

Suppose that at time 0 the particle is at $z = 0$ and at time $t_1$ it is at $z = z_1$. For any motion $z(t)$, actual or virtual, between these two points the action is

$$S[z(t)] = \int_0^{t_1} L(z,\dot{z})\,dt.$$

Pretend you don't know what the actual motion is. You might then guess that it can be adequately represented by the first three terms in a power series in t,

$$z = z_0 + v_0 t + \tfrac{1}{2}at^2,$$

where $z_0$ and $v_0$ are chosen so that $z(t)$ passes through the end points, and a is an adjustable parameter. Evaluate S for this form of $z(t)$ and note the dependence of S on a. For what value of a is S a minimum?

# CHAPTER V

# INVARIANCE TRANSFORMATIONS AND CONSTANTS OF THE MOTION

In this chapter we explore the remarkable connection between **symmetry**, the invariance of a system under transformations, and **conservation laws**, the existence of constants of the motion. This relation usually goes by the name **Noether's theorem.**[1] Of special importance is the symmetry of any closed system under the space-time transformations, which allows us to infer the conservation of linear momentum, angular momentum, and energy, and the center of mass theorem -- this without knowing the detailed equations of motion for the system.

## Invariance transformations

One of the great advantages of Lagrangian dynamics is the freedom it allows in the choice of generalized coordinates. If $q_a$ is a set of coordinates, then any reversible **point transformation**

$$q'_a = q'_a(q_1,\cdots,q_f;t)$$

defines another set $q'_a$. This new set satisfies Lagrange's equations of motion with new Lagrangian

$$L'(q',\dot{q}',t) = L(q(q',t),\dot{q}(q',\dot{q}',t),t).$$

This equation states: to obtain the Lagrangian $L'$ for the new coordinates, use the inverse transformation equations to express the old coordinates and their time derivatives in the old Lagrangian L in terms of the new coordinates and their time derivatives; the resulting function of $q'$, $q'$, and t is the new Lagrangian $L'$.

Although the general form of Lagrange's equations of motion is preserved in any point transformation, the explicit equations of motion for the new variables usually look different from those for the old variables; they cannot be obtained simply by replacing old variables by new. However, for a given system there may be particular transformations for which the explicit equations of motion *are* the same for the old and new variables. The equations of motion (or the system) are then said to be **invariant** under these transformations, and such transformations are called **invariance transformations**. A transformation is certainly an invariance transformation if the Lagrangian itself is invariant,

$$L'(q',\dot{q}',t) = L(q',\dot{q}',t);$$

---

[1]E. Noether, "Invariante variationsprobleme," Nachr. Akad. Wiss. Göttingen, Math.-Phys. Kl. II **1918**, 235-257 (1918); an English translation by M. A. Tavel is available in Transp. Theory Stat. Phys. **1**, 186-207 (1971); a good review is E. L. Hill, "Hamilton's Principle and the Conservation Theorems of Mathematical Physics," Rev. Mod. Phys. **23**, 253-260 (1951).

that is, if the new Lagrangian is the same function of the new variables as the old was of the old. This, however, is too restrictive. If we only require that $L'$ and $L$ lead to the same explicit equations of motion, this is still the case if $L'(q',\dot{q}',t)$ differs from $L(q',\dot{q}',t)$ by a term which gives zero identically (independent of $q'(t)$) when substituted into Lagrange's equations. Such a term has the form

$$\frac{d\Lambda(q',t)}{dt} = \sum_{a=1}^{f} \frac{\partial \Lambda}{\partial q'_a}\dot{q}'_a + \frac{\partial \Lambda}{\partial t}.$$

The condition for invariance is thus

$$L'(q',\dot{q}',t) = L(q',\dot{q}',t) + \frac{d\Lambda(q',t)}{dt}$$

where $\Lambda$ is some function of $q'$ and $t$ which is determined by the transformation. Combining this with the definition of $L'$, we can write the condition for invariance in the form

$$L(q,\dot{q},t) = L(q',\dot{q}',t) + \frac{d\Lambda(q',t)}{dt}.$$

We can use this to check whether or not a given transformation is an invariance transformation.

## Free particle (a)

The motion in one dimension x of a free particle of mass m is described by the Lagrangian

$$L = \tfrac{1}{2}m\dot{x}^2.$$

Let us consider the possible invariance of this system under a transformation of the form

$$x' = x + a \qquad \dot{x}' = \dot{x} + \dot{a}$$

where a may be some function of time. We construct the Lagrangian $L'$ for the new coordinate $x'$,

$$L'(x',\dot{x}',t) = \tfrac{1}{2}m(\dot{x}' - \dot{a})^2 = L(x',\dot{x}',t) - m\dot{a}\dot{x}' + \tfrac{1}{2}m\dot{a}^2.$$

In order for this transformation to be an invariance transformation, the last two terms must be the total time derivative of some function $\Lambda(x',t)$. This requires that

$$\frac{\partial\Lambda}{\partial x'} = -m\dot{a} \qquad \text{and} \qquad \frac{\partial\Lambda}{\partial t} = \tfrac{1}{2}m\dot{a}^2.$$

Together these imply $\ddot{a} = 0$, which in turn gives

$$a = \alpha + \beta t \qquad \Lambda = -m\beta x' + \tfrac{1}{2}m\beta^2 t$$

where $\alpha$ and $\beta$ are arbitrary constants. The most general invariance transformation of the above form is thus

$$x' = x + \alpha + \beta t$$

and consists of spatial displacement for $\beta = 0$, and **Galilean transformation** (transformation to a coordinate system moving with a uniform velocity with respect to the original system) for $\beta \neq 0$. These are the well-known and expected invariances for a free particle. It is worth noting that a transformation to a uniformly accelerating frame is *not* an invariance transformation. This is also clear from Newton's second law, since in an accelerated frame there are so-called inertial forces (not due to the presence of nearby matter) which are not present in an inertial frame.

## Infinitesimal transformations

Many transformations, such as the one just discussed, contain adjustable parameters "$\alpha$." Further, they are such that for particular values $\alpha^0$ of the parameters the transformation reduces to the identity, thus

$$q'_a = q'_a(q;\alpha;t) \qquad \text{with} \qquad q'_a(q;\alpha^0;t) = q_a.$$

Transformations for which the parameters are infinitesimally close to $\alpha^0$, and for which the new coordinates $q'_a$ differ infinitesimally from the old coordinates $q_a$, are called **infinitesimal transformations**. For these we have

$$\alpha_v = \alpha_v^0 + \delta\alpha_v \quad \text{and} \quad q'_a = q_a + \delta q_a \quad \text{where} \quad \delta q_a = \sum_v \left(\frac{\partial q'_a}{\partial \alpha_v}\right)_0 \delta\alpha_v.$$

Of special interest are the **infinitesimal invariance transformations**, for which

$$L(q,\dot{q},t) = L(q + \delta q, \dot{q} + \delta\dot{q}, t) + \frac{d\delta\Lambda}{dt}.$$

Expanding this to first order in the small quantities $\delta q$ and $\delta\dot{q}$ and rearranging the result, we obtain

$$\frac{d}{dt}\left(\sum_{a=1}^{f}\frac{\partial L}{\partial \dot{q}_a}\delta q_a + \delta\Lambda\right) + \sum_{a=1}^{f}\left[\frac{\partial L}{\partial q_a} - \frac{d}{dt}\left(\frac{\partial L}{\partial \dot{q}_a}\right)\right]\delta q_a = 0.$$

The second term vanishes if the generalized coordinates $q_a$ satisfy Lagrange's equations of motion, and therefore the quantity

$$\sum_{a=1}^{f}\frac{\partial L}{\partial \dot{q}_a}\delta q_a + \delta\Lambda$$

is a **constant of the motion**. This combination of the generalized coordinates $q_a$, the generalized velocities $\dot{q}_a$, and the time t remains constant as the system develops in time. We thus have the important result:

Associated with
any infinitesimal invariance transformation
is
a constant of the motion.

The fact that the generalized momentum associated with a cyclic coordinate is a constant of the motion, which we noted earlier, is a special case of this result. For if $q_a$ is a cyclic coordinate, it does not appear in the Lagrangian, and the Lagrangian is invariant under the transformation $q_a' = q_a + \alpha$ where $\alpha$ is an arbitrary constant. The constant of the motion associated with the corresponding infinitesimal transformation is $(\partial L/\partial \dot{q}_a)\delta\alpha$, the generalized momentum (times $\delta\alpha$).

## Free particle (b)

We have seen that

$$x' = x + \alpha + \beta t$$

is an invariance transformation for a free particle in one dimension. It depends on the two parameters $\alpha$ and $\beta$ and reduces to the identity transformation if the parameters are zero. The corresponding infinitesimal transformation is

$$x' - x = \delta x = \delta\alpha + t\,\delta\beta.$$

The new and old Lagrangians differ by $d(\delta\Lambda)/dt$ where

$$\delta\Lambda = -mx\,\delta\beta.$$

To obtain $\delta\Lambda$ from the $\Lambda$ given earlier, we have replaced $\beta$ by $\delta\beta$ and retained only first order terms in $\delta\beta$; we have also replaced $x'$ by $x$, the difference between these contributing only (negligible) second order terms to $\delta\Lambda$.

Associated with this infinitesimal invariance transformation is the constant of the motion

$$m\dot{x}(\delta\alpha + t\,\delta\beta) - mx\,\delta\beta = m\dot{x}\,\delta\alpha + (m\dot{x}t - mx)\delta\beta .$$

Associated with spatial displacements is linear momentum $m\dot{x}$, and associated with Galilean transformations is $m\dot{x}t - mx$; this second constant is $-m$ times the initial position of the particle.

## Space-time transformations

Newtonian space is homogeneous (the same at all points) and isotropic (the same in all directions), and Newtonian time is homogeneous. Further, Newtonian space remains unchanged under Galilean transformations (transformations between frames moving with uniform velocity with respect to one another). Thus a closed system (one not interacting with other systems) behaves the same no matter where it is located, how it is oriented, when it is started, or whether or not it has a uniform velocity. It is invariant under the space-time transformations of spatial displacement, spatial rotation, time displacement, and Galilean transformation. These transformations can be looked at from one of two perspectives:

(a) the **passive** point of view in which we imagine describing the system from two frames of reference which are displaced, rotated, etc. relative to one another (or imagine two such observers). The system remains unchanged. Up to this point, we have been using the passive interpretation.

(b) the **active** point of view in which we imagine displacing, rotating, etc., the system (or alternatively, we compare two identical systems which are displaced, rotated, etc. relative to one another). The observer remains unchanged. The mathematical formalism is the same, regardless of the perspective.

Associated with each space-time invariance is a constant of the motion. These constants are of special importance, since they apply to any closed system. Further, if the system can be split into two non-interacting parts, the constant for the system is the sum of the constants for the individual parts, since the Lagrangian for the system is the sum of the individual Lagrangians. The constants are additive. These constants are of great use in analyzing processes, such as collisions, for which we have incomplete knowledge. As a specific example we consider a closed system of N particles interacting with one another via potentials which depend only on the distances between the particles. The Lagrangian is

$$L = \sum_{i=1}^{N} \tfrac{1}{2} m_i |\dot{\mathbf{r}}_i|^2 - \tfrac{1}{2} \sum_{\substack{i=1\\ i\neq j}}^{N} \sum_{j=1}^{N} V_{ij}(|\mathbf{r}_i - \mathbf{r}_j|) .$$

## Spatial displacement

Suppose we view the system from a new coordinate frame which is displaced an amount $-\mathbf{a}$ with respect to the original frame (Fig. 5.01(a)). Alternatively, we displace the system an amount $\mathbf{a}$ (Fig. 5.01(b)).

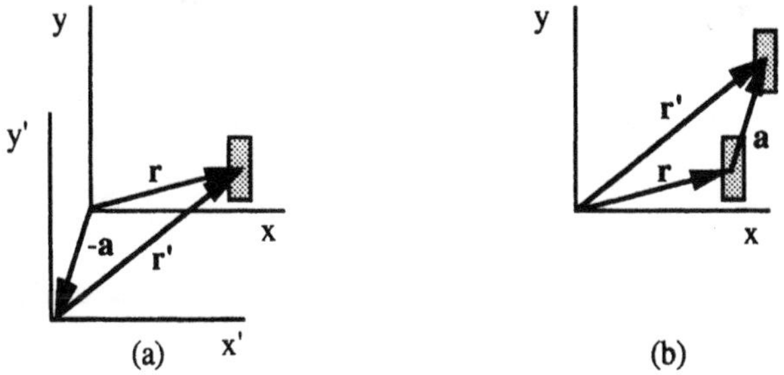


Fig. 5.01. Spatial displacement (a) passive, (b) active

From either point of view we have

$$\mathbf{r}_i' = \mathbf{r}_i + \mathbf{a} \qquad \dot{\mathbf{r}}_i' = \dot{\mathbf{r}}_i .$$

The Lagrangian itself is invariant under this transformation. The constant of the motion associated with the corresponding infinitesimal transformation is

$$\sum_{i=1}^{N} \frac{\partial L}{\partial \dot{\mathbf{r}}_i} \cdot \delta\mathbf{a} = \sum_{i=1}^{N} m_i \dot{\mathbf{r}}_i \cdot \delta\mathbf{a} = \mathbf{P} \cdot \delta\mathbf{a}$$

where $\mathbf{P} = \sum_{i=1}^{N} m_i \dot{\mathbf{r}}_i$ is the total **linear momentum** of the system. Note that we reach this conclusion without knowing the specific form of the interaction $V_{ij}(|\mathbf{r}_i - \mathbf{r}_j|)$.

## Spatial rotation

Suppose we view the system from a new coordinate frame which is rotated with respect to the original frame through an angle $-\theta$ about the z-axis (Fig. 5.02(a)). Alternatively, we rotate the system through an angle $\theta$ about the z-axis (Fig. 5.02(b)).

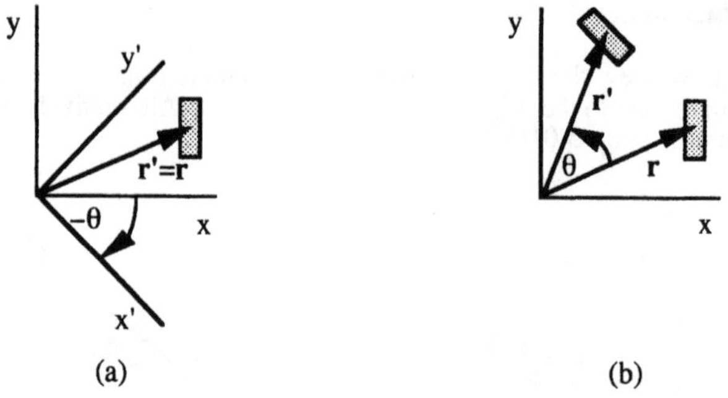


Fig. 5.02. Spatial rotation (a) passive, (b) active

The coordinates are related by

$$x' = x\cos\theta - y\sin\theta$$
$$y' = x\sin\theta + y\cos\theta$$
$$z' = z \;.$$

The Lagrangian itself is invariant under this transformation. The corresponding infinitesimal transformation is obtained by replacing $\theta$ by a "small" $\delta\theta$ and noting that $\cos\delta\theta \approx 1$, $\sin\delta\theta \approx \delta\theta$; thus

$$x' - x = \delta x = -y\,\delta\theta$$
$$y' - y = \delta y = x\,\delta\theta$$
$$z' = z \;.$$

The associated constant of the motion is

$$\sum_{i=1}^{N}\left(-\frac{\partial L}{\partial \dot{x}_i} y_i\,\delta\theta + \frac{\partial L}{\partial \dot{y}_i} x_i\,\delta\theta\right) = \sum_{i=1}^{N} m_i\left(-y_i\dot{x}_i + x_i\dot{y}_i\right)\delta\theta = L_z\,\delta\theta$$

where $L_z$ is the z-component of the total angular momentum. Similarly, invariance under rotations about the x and y axes leads to constancy of the x and y components of the total angular momentum.

Indeed, we can consider all three components together, specifying an infinitesimal rotation by a vector $\boldsymbol{\delta\theta}$, which represents a rotation through an angle $|\boldsymbol{\delta\theta}|$ about an axis $\boldsymbol{\delta\theta}/|\boldsymbol{\delta\theta}|$.[2] See Fig. 5.03.

---

[2]This does not work for finite rotations, since a finite "rotation" $\boldsymbol{\theta}_1$ and a finite "rotation" $\boldsymbol{\theta}_2$ do not add like vectors.

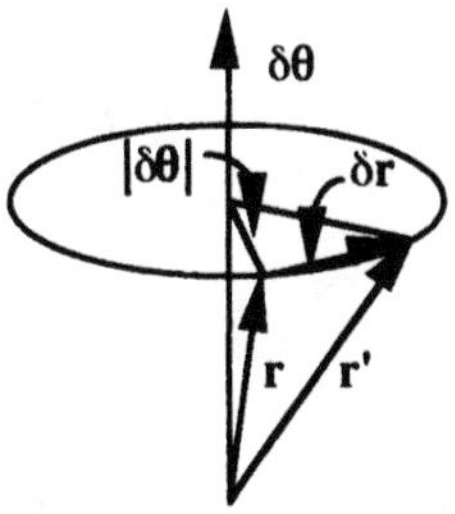


Fig. 5.03. Infinitesimal rotation

We have

$$\mathbf{r}' - \mathbf{r} = \delta\mathbf{r} = \delta\boldsymbol{\theta} \times \mathbf{r},$$

and the constant of the motion associated with this infinitesimal invariance transformation is

$$\sum_{i=1}^{N} \frac{\partial L}{\partial \dot{\mathbf{r}}_i} \cdot \delta\mathbf{r}_i = \sum_{i=1}^{N} m_i \dot{\mathbf{r}}_i \cdot \delta\boldsymbol{\theta} \times \mathbf{r}_i = \mathbf{L} \cdot \delta\boldsymbol{\theta}$$

where $\mathbf{L} = \sum_{i=1}^{N} \mathbf{r}_i \times m_i \dot{\mathbf{r}}_i$ is the total **angular momentum** of the system.

## Galilean transformation

Suppose we view the system from a new coordinate frame which is moving with a uniform velocity $-\mathbf{v}$ with respect to the original frame (Fig. 5.04(a)). Alternatively, we give each of the particles of the system an additional velocity $\mathbf{v}$ (Fig. 5.04(b)).

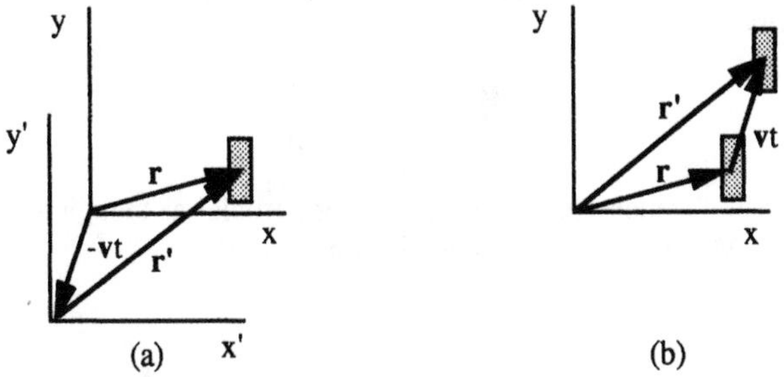


Fig. 5.04. Galilean transformation (a) passive, (b) active

The coordinates are related by

$$\mathbf{r}_i' = \mathbf{r}_i + \mathbf{v}t \qquad \dot{\mathbf{r}}_i' = \dot{\mathbf{r}}_i + \mathbf{v}.$$

The new Lagrangian is given by

$$L'(q',\dot{q}',t) = \sum_{i=1}^{N} \tfrac{1}{2} m_i |\dot{\mathbf{r}}_i' - \mathbf{v}|^2 - \tfrac{1}{2} \sum_{i=1}^{N} \sum_{j=1}^{N} V_{ij}(|\mathbf{r}_i' - \mathbf{v}t - \mathbf{r}_j' + \mathbf{v}t|)$$

$$= L(q',\dot{q}') - \sum_{i=1}^{N} m_i \dot{\mathbf{r}}_i' \cdot \mathbf{v} + \sum_{i=1}^{N} \tfrac{1}{2} m_i |\mathbf{v}|^2 \; .$$

In this case the Lagrangian is *not* invariant; however, the equations of motion are, since the last two terms can be written as the total time derivative of the function

$$\Lambda = -\sum_{i=1}^{N} m_i \mathbf{r}_i' \cdot \mathbf{v} + \sum_{i=1}^{N} \tfrac{1}{2} m_i |\mathbf{v}|^2 \, t.$$

The constant of the motion associated with the infinitesimal invariance transformation is

$$\sum_{i=1}^{N} \frac{\partial L}{\partial \dot{\mathbf{r}}_i} \cdot \delta \mathbf{r}_i + \delta\Lambda = \sum_{i=1}^{N} (m_i \dot{\mathbf{r}}_i t - m_i \mathbf{r}_i) \cdot \delta \mathbf{v} = [\mathbf{P}t - M\mathbf{R}_{cm}] \cdot \delta \mathbf{v}$$

where $M = \sum_{i=1}^{N} m_i$ is the total mass and $\mathbf{R}_{cm} = \sum_{i=1}^{N} m_i \mathbf{r}_i \,/\, M$ is the radius vector to the center of mass of the system. We thus have

$$\mathbf{R}_{cm}(t) = \mathbf{R}_{cm}(0) + (\mathbf{P}/M)t,$$

where we have written the "constant of the motion" as $-M\mathbf{R}_{cm}(0)$. The quantity $\mathbf{R}_{cm}(0)$ can be identified as the radius vector to the initial $(t = 0)$ location of the center of mass. This is the **center of mass theorem**: the center of mass of a closed system moves with a constant velocity

$$\frac{d\mathbf{R}_{cm}}{dt} = \mathbf{V}_{cm} = \frac{\mathbf{P}}{M}.$$

It allows us to treat, for many purposes, systems of particles as though they were a single particle with mass M, location $\mathbf{R}_{cm}$, and momentum $\mathbf{P} = M\mathbf{V}_{cm}$.

### Time displacement

We have not yet considered invariance under time displacement. The reason is that this transformation involves changes to the *independent* variable t, and the formalism we have developed to this point is only capable of handling changes to the *dependent* variables $q_a$. We shall make the appropriate generalizations in the next section, but since the considerations there are rather lengthy, we give here a brief independent discussion applicable to this particular case. Let us consider the (total) time rate of change of the Lagrangian,

$$\frac{dL}{dt} = \sum_{a=1}^{f} \frac{\partial L}{\partial q_a}\dot{q}_a + \sum_{a=1}^{f} \frac{\partial L}{\partial \dot{q}_a}\frac{d\dot{q}_a}{dt} + \frac{\partial L}{\partial t}.$$

The second term can be rewritten

$$\frac{d}{dt}\left(\sum_{a=1}^{f} \frac{\partial L}{\partial \dot{q}_a}\dot{q}_a\right) - \frac{d}{dt}\left(\sum_{a=1}^{f} \frac{\partial L}{\partial \dot{q}_a}\right)\dot{q}_a,$$

and thus the equation for the time rate of change of the Lagrangian can be rearranged in the form

$$\frac{d}{dt}\left(\sum_{a=1}^{f} \frac{\partial L}{\partial \dot{q}_a}\dot{q}_a - L\right) = -\sum_{a=1}^{f}\left(\frac{\partial L}{\partial q_a} - \frac{d}{dt}\left(\frac{\partial L}{\partial \dot{q}_a}\right)\right)\dot{q}_a - \frac{\partial L}{\partial t}.$$

Now if the generalized coordinates $q_a$ satisfy Lagrange's equations, the first term on the right is zero, and we are left with

$$\frac{dH}{dt} = -\frac{\partial L}{\partial t},$$

where

$$H = \sum_{a=1}^{f} \frac{\partial L}{\partial \dot{q}_a}\dot{q}_a - L$$

is called the **Hamiltonian**.[3] If the Lagrangian is the difference of the kinetic and potential energies, $L = T - V$, and if the kinetic energy is a homogeneous quadratic function of the generalized velocities, $T = \sum_{a=1}^{f}\sum_{b=1}^{f} A_{ab}\dot{q}_a\dot{q}_b$, then $\sum_{a=1}^{f}\left(\partial L/\partial \dot{q}_a\right)\dot{q}_a = 2T$ and

$$H = 2T - (T - V) = T + V.$$

In this case H is the total energy. In any event, it may be thought of as a "generalized energy." The above equation shows that if the Lagrangian does not depend explicitly on time, so that the equations of motion are invariant under time displacement, then the Hamiltonian is constant in time.

---

[3]Strictly speaking, the Hamiltonian should be expressed in terms of the appropriate variables, which turn out to be the generalized coordinates and *momenta*, as we discuss more fully in the next chapter.

## Covariance, invariance, and the action

The principle of stationary action summarizes the dynamical behavior of a system in a compact and elegant way. It thus provides a convenient foundation on which to base a general discussion of the ideas of covariance, invariance, and their connection with constants of the motion. We give here a self-contained account of these aspects of classical mechanics from this unified point of view, even though many of the ideas have already been discussed, at least for special cases. In order to avoid unnecessary clutter, we consider explicitly a system with one degree of freedom. The generalization to many degrees is trivial; all we must do is to insert a few subscripts and summation signs.

We are interested in the effect on paths C in configuration space, and on the action S[C] for such paths, as a result of subjecting configuration space to an **extended point transformation**

$$q' = q'(q,t) \qquad t' = t'(q,t)$$

involving both the dependent variable, the generalized coordinate q, and the independent variable, the time t (Fig. 5.05).

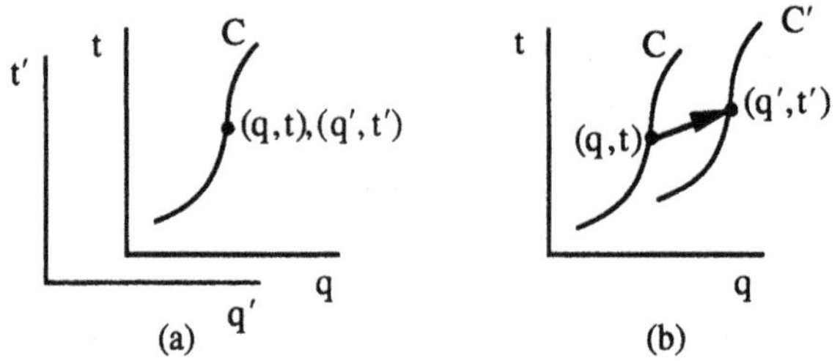


Fig. 5.05. Transformation of configuration space (a) passive, (b) active

As with the linear space-time transformations, we can interpret this in one of two ways:

(a) the *passive* point of view in which we picture $(q',t')$ and $(q,t)$ as the coordinates assigned to a single physical point by two observers, "prime" and "unprime." From this point of view, the transformation amounts simply to a change of the variables used to describe the system.

(b) the *active* point of view in which we picture system points $(q,t)$ as being moved to new locations $(q',t')$, and hence system paths C as changed into new paths C'. The cbserver remains unchanged.

It is convenient to begin by adopting the passive point of view. Let us suppose that path C is a possible actual motion of the system. The action

$$S[C] = \int_{t_0}^{t_1} L(q, dq/dt, t)\,dt$$

for C, as expressed by the unprime observer, is stationary when compared with that for neighboring paths with the same end points, and C is given as a solution to Lagrange's equation

$$\frac{d}{dt}\left(\frac{\partial L}{\partial(dq/dt)}\right)=\frac{\partial L}{\partial q}.$$

For the prime observer the path C is still a possible actual path; the action has the same value as for the unprime observer and is stationary. We are simply changing variables. However, the prime observer writes

$$S[C]=\int_{t'_0}^{t'_1} L'(q',dq'/dt',t')dt'$$

with

$$L'(q',dq'/dt',t')dt'=L(q,dq/dt,t)dt.$$

This equation states: use the transformation equations to express the unprime variables on the right-hand side in terms of the prime variables; this gives the Lagrangian $L'=(dt/dt')L$ as used by the prime observer. In general $L'$ has, as a function of $q'$, $\dot{q}'$, and $t'$, a different functional form than L has of $q$, $\dot{q}$, and t. Further, because of the factor $(dt/dt')$, its value at a given physical point is, in general, not the same for the two observers. To describe the path C, the prime observer writes

$$\frac{d}{dt'}\left(\frac{\partial L'}{\partial(dq'/dt')}\right)=\frac{\partial L'}{\partial q'},$$

again Lagrange's equation, with new variables and a new Lagrangian $L'$. We say that Lagrange's equation is **covariant**, maintains the same general form, under arbitrary transformations of the dependent and independent variables. This is a generalization of our earlier results, which considered transformations of the dependent variables alone.

The explicit equation of motion written down by the prime observer does not in general look the same as that written down by the unprime observer; it cannot be obtained simply by replacing unprime variables by prime. For particular transformations, however, it may turn out that the equations *do* have the same explicit form. If this is so, we say that the transformation is an **invariance** transformation. The observers are then equivalent. Our previous considerations show that we have invariance if

$$L'(q',dq'/dt',t')=L(q',dq'/dt',t')+d\Lambda/dt'.$$

The consequences of invariance can now be worked out by combining this with the above definition of $L'$ and proceeding along lines similar to those at the beginning of the chapter.

It is, however, more interesting to approach invariance by considering the action and what happens to it under an extended point transformation, now looked at from the active point of view. The transformation

$$q'=q'(q,t) \qquad t'=t'(q,t)$$

from the active point of view carries system points into new system points and system paths into new paths. *If* the transformation carries possible actual paths into other possible actual paths, we say that it is an invariance transformation. Let us suppose that C is an actual path and that C′, the path obtained from C by active transformation, is also a possible actual path. Path C has the property that the action for it

$$S[C] = \int_{t_0}^{t_1} L(q, dq/dt, t)\,dt$$

is stationary when compared with that for neighboring paths with the same end points. For invariance, the action for path C′

$$S[C'] = \int_{t_0'}^{t_1'} L(q', dq'/dt', t')\,dt'$$

must be stationary when compared with that for neighboring paths to *it* (note that the Lagrangian is L; there is only one observer, looking at two different paths). S[C′] can thus differ from S[C] by at most a term which gives zero under all fixed end point variations. Such a term can be a function only of the end points, and the additive property

$$S[0 \rightarrow 2] = S[0 \rightarrow 1] + S[1 \rightarrow 2]$$

of the action further restricts this function to the form

$$-[\Lambda(q_1', t_1') - \Lambda(q_0', t_0')] = -[\Lambda]_{t_0'}^{t_1'}.$$

(Compare the argument used in elementary mechanics in introducing the potential energy.) Invariance thus requires that the difference of the actions

$$\Delta S = S[C'] - S[C] = \int_{t_0'}^{t_1'} L(q', dq'/dt', t')\,dt' - \int_{t_0}^{t_1} L(q, dq/dt, t)\,dt$$

for the original and transformed paths C and C′ be given by

$$\Delta S = -[\Lambda]_{t_0'}^{t_1'}.$$

This may be taken as the condition for invariance, now expressed in terms of the action.

To continue the discussion, let us now consider two arbitrary neighboring paths with corresponding points connected by the infinitesimal transformation (Fig. 5.06)

$$q' = q + \Delta q(q, t) \qquad t' = t + \Delta t(q, t)$$

where $\Delta q$ and $\Delta t$ are "small," and let us work out the difference in the action for the two paths. We use $\Delta$ rather than $\delta$ to denote the infinitesimal changes, since we wish to reserve $\delta$ for a virtual (time-frozen) change.

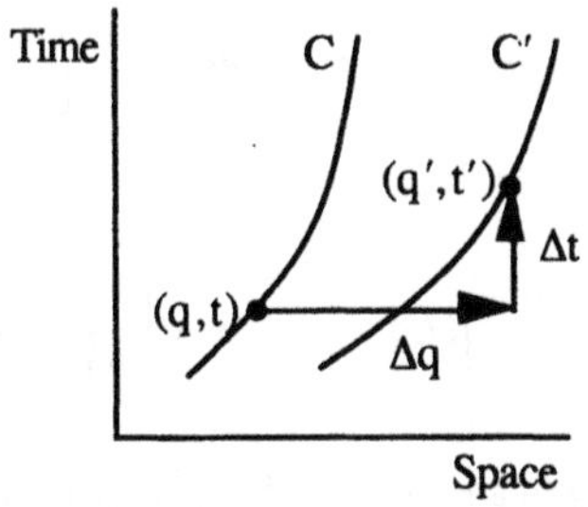


Fig. 5.06. Infinitesimal transformation of configuration space

We need

$$dq' = dq + (d\Delta q/dt)dt \quad \text{and} \quad dt' = (1 + (d\Delta t/dt))dt.$$

(note the total time derivative $(d/dt) = (\partial/\partial t) + \dot{q}(\partial/\partial q)$). Together these yield the relation between the generalized velocity at corresponding points on the original and transformed paths

$$\frac{dq'}{dt'} = \left(1 + \frac{d\Delta t}{dt}\right)^{-1}\left(\frac{dq}{dt} + \frac{d\Delta q}{dt}\right) \approx \left(1 - \frac{d\Delta t}{dt}\right)\frac{dq}{dt} + \frac{d\Delta q}{dt}.$$

Note that the change in the generalized velocity is no longer simply the time derivative of the change in the generalized coordinate. The difference in the action for the two neighboring paths now follows

$$\Delta S = \int_{t_0}^{t_1}\left[L\left(q + \Delta q, \left(1 - \frac{d\Delta t}{dt}\right)\frac{dq}{dt} + \frac{d\Delta q}{dt}, t + \Delta t\right)\left(1 + \frac{d\Delta t}{dt}\right) - L\left(q, \frac{dq}{dt}, t\right)\right]dt.$$

Expanding the integrand via Taylor's theorem, we obtain to first order in the small quantities $\Delta q$ and $\Delta t$

$$\Delta S = \int_{t_0}^{t_1}\left[\frac{\partial L}{\partial q}\Delta q + \frac{\partial L}{\partial \dot{q}}\left(\frac{d\Delta q}{dt} - \dot{q}\frac{d\Delta t}{dt}\right) + \frac{\partial L}{\partial t}\Delta t + L\frac{d\Delta t}{dt}\right]dt.$$

Integrating the second and fourth terms by parts, we then get

$$\Delta S = \int_{t_0}^{t_1}\left[\left(\frac{\partial L}{\partial q} - \frac{d}{dt}\left(\frac{\partial L}{\partial \dot{q}}\right)\right)\Delta q + \left(\frac{d}{dt}\left(\frac{\partial L}{\partial \dot{q}}\dot{q}\right) + \frac{\partial L}{\partial t} - \frac{dL}{dt}\right)\Delta t\right]dt + \left[\frac{\partial L}{\partial \dot{q}}\Delta q - \left(\frac{\partial L}{\partial \dot{q}}\dot{q} - L\right)\Delta t\right]_{t_0}^{t_1}$$

The coefficient of $\Delta t$ in the integrand can be written $-\left(\frac{\partial L}{\partial q} - \frac{d}{dt}\left(\frac{\partial L}{\partial \dot{q}}\right)\right)\dot{q}\Delta t$, so we finally find

$$\Delta S = \int_{t_0}^{t_1} \left[\left(\frac{\partial L}{\partial q} - \frac{d}{dt}\left(\frac{\partial L}{\partial \dot{q}}\right)\right)(\Delta q - \dot{q}\Delta t)\right] dt + \left[\frac{\partial L}{\partial \dot{q}}\Delta q - \left(\frac{\partial L}{\partial \dot{q}}\dot{q} - L\right)\Delta t\right]_{t_0}^{t_1} .$$

This result is general, applying to any two neighboring paths in configuration space. If the original path is a possible actual path satisfying Lagrange's equation, the integrand is zero and the expression for $\Delta S$ simplifies to

$$\Delta S = \left[\frac{\partial L}{\partial \dot{q}}\Delta q - \left(\frac{\partial L}{\partial \dot{q}}\dot{q} - L\right)\Delta t\right]_{t_0}^{t_1} .$$

We can simplify this still further by noticing that the coefficient of $\Delta q$ is the generalized momentum p, and the coefficient of $\Delta t$ is the Hamiltonian H, so we end up with the compact and beautiful result

$$\Delta S = \left[p\Delta q - H\Delta t\right]_{t_0}^{t_1} .$$

This result has many uses, and we shall return to it again. For the present we note that if the transformation is an infinitesimal *invariance* transformation, we can combine the result with the condition $\Delta S = -[\Delta\Lambda]_{t_0}^{t_1}$ to find

$$\left[p\Delta q - H\Delta t + \Delta\Lambda\right]_{t_0}^{t_1} = 0.$$

That is, the quantity in square brackets is constant in time, and we can say:

Associated with the infinitesimal invariance transformation
$q' = q + \Delta q(q,t) \qquad t' = t + \Delta t(q,t)$
is the constant of the motion
$p\Delta q - H\Delta t + \Delta\Lambda$.

This is the generalization of our previous result to include transformations which change the independent variable t. We can use it to derive again the connection between invariance under time displacement and conservation of energy. If the Lagrangian does not depend explicitly on the time, the system is invariant (with $\Lambda = 0$) under time displacement

$$q' = q \qquad t' = t + \Delta t,$$

where $\Delta t$ is any constant. The associated constant of the motion is $-H\Delta t$, "generalized energy" or "Hamiltonian" (times $-\Delta t$).

We have now achieved our main objective of seeing the connection between invariance transformations and constants of the motion from the point of view of action. However, the result for the difference in the action for two neighboring paths, say C and C′, is sufficiently interesting that it is worthwhile to give a different, perhaps simpler, derivation. This time let us use the *same* time variable t for both paths. Corresponding points on C and C′ are now at the same time (so this is now a virtual displacement) and their generalized coordinates are related by (Fig. 5.07(a))

$$q' = q + \delta q(q,t).$$

Note that for given q and t this is not the same infinitesimal spatial displacement as before. The relation between it and the earlier space and time displacements $\Delta q$ and $\Delta t$ can be seen from Fig. 5.07(b), thus

$$\delta q = \Delta q - \dot{q}\Delta t.$$

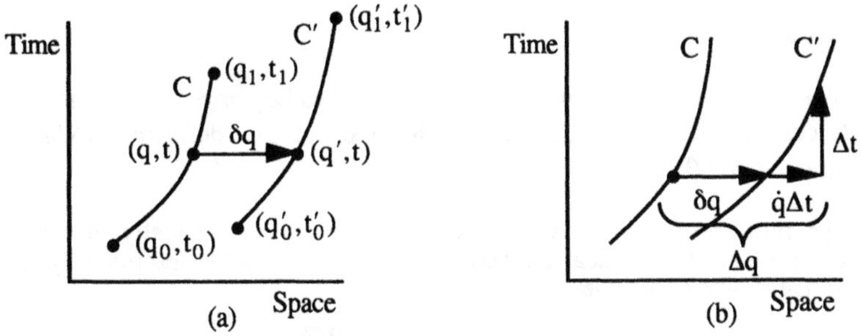


Fig. 5.07. Infinitesimal transformation of configuration space
(a) viewed as a virtual displacement
(b) relation to previous point of view

The difference in the action for the two paths is now given by

$$\Delta S = \int_{t_0'}^{t_1'} L(q', dq'/dt, t)\,dt - \int_{t_0}^{t_1} L(q, dq/dt, t)\,dt.$$

The first integral can be broken up into three parts, $\int_{t_0'}^{t_1'} = \int_{t_0}^{t_1} + \int_{t_1}^{t_1'} - \int_{t_0}^{t_0'}$, the range of the first part matching that of the second integral, and the last two parts arising from the (infinitesimal) shifts in the limits. For these the integral is simply the Lagrangian L times the shift $\Delta t = t' - t$, so we have

$$\Delta S = \int_{t_0}^{t_1}[L(q',dq'/dt,t) - L(q,dq/dt,t)]dt + [L\Delta t]_{t_0}^{t_1}.$$

The integrand can be written (compare our earlier discussion)

$$L\left(q+\delta q, \frac{dq}{dt}+\frac{d\delta q}{dt}, t\right) - L\left(q,\frac{dq}{dt},t\right) = \frac{\partial L}{\partial q}\delta q + \frac{\partial L}{\partial \dot{q}}\frac{d\delta q}{dt} = \left[\frac{\partial L}{\partial q} - \frac{d}{dt}\left(\frac{\partial L}{\partial \dot{q}}\right)\right]\delta q + \frac{d}{dt}\left(\frac{\partial L}{\partial \dot{q}}\delta q\right)$$

The first term is zero if q is an actual path, and the second term is a total time derivative, so we find

$$\Delta S = \left[\frac{\partial L}{\partial \dot{q}}\delta q + L\Delta t\right]_{t_0}^{t_1}.$$

Finally, recalling that $\delta q = \Delta q - \dot{q}\Delta t$, we recover the relation $\Delta S = [p\Delta q - H\Delta t]_{t_0}^{t_1}$.

## Exercises

1. Show that a function of $q(t)$, $\dot{q}(t)$, and t satisfies Lagrange's equations identically (independent of $q_a(t)$) if, and only if, it is the total time derivative $d\Lambda/dt$ of some function $\Lambda(q(t),t)$.

2. The motion of a particle of mass m which moves vertically in the uniform gravitational field g near the surface of the earth can be described by an action principle with Lagrangian
$$L = \tfrac{1}{2}m\dot{z}^2 - mgz.$$
(a) Show that the action principle is invariant under the transformations
(i) $z' = z + \alpha$ where $\alpha$ is any constant;
(ii) $z' = z + \beta t$ where $\beta$ is any constant.
(b) Find the associated constants of the motion.

3. The motion of a simple harmonic oscillator is described by an action principle with Lagrangian
$$L = \tfrac{1}{2}m\dot{x}^2 - \tfrac{1}{2}m\omega^2x^2$$
(a) Show that the action principle is invariant under the two-parameter $(A,B)$ family of transformations
$$x' = x + A\sin\omega t + B\cos\omega t.$$
(b) Find the two independent constants of the motion associated with the infinitesimal transformation, and identify them physically.
(c) Use the results of (b) to write down the general solution to the equation of motion.

4. The Lagrangian for a particle of mass m and charge e moving in a uniform magnetic field which points in the z-direction is (see exercise 3.14)

$$L = \frac{1}{2}m(\dot{x}^2 + \dot{y}^2 + \dot{z}^2) + (eB/2c)(x\dot{y} - y\dot{x}).$$

(a) Show that the system is invariant under spatial displacement (in any direction) and find the associated constants of the motion.
(b) Show that the system is invariant under rotation about the z-axis and find the associated constant of the motion.

# CHAPTER VI

# HAMILTON'S EQUATIONS

Our studies in mechanics to this point have been largely based on Lagrange's equations. In this chapter we replace Lagrange's equations, which are a set of second order differential equations, by an equivalent set of first order equations, Hamilton's equations. The simple form and high degree of structure of these latter equations makes them most suitable for our subsequent studies in advanced mechanics.

## Hamilton's equations

In Lagrangian mechanics a system of f degrees of freedom is described by f generalized coordinates $(q_1,q_2,\ldots,q_f)$ which satisfy Lagrange's equations

$$\frac{d}{dt}\left(\frac{\partial L}{\partial \dot{q}_a}\right) = \frac{\partial L}{\partial q_a},$$

a set of f second order differential equations for the coordinates. Lagrange's equations, or indeed any set of f second order equations, can be replaced by an equivalent set of 2f first order equations. There are many ways in which this can be done. One way is to introduce as new independent variables the generalized velocities $v_a = \dot{q}_a$. Lagrange's equations then become

$$\frac{d}{dt}\left(\frac{\partial L(q,v,t)}{\partial v_a}\right) = \frac{\partial L(q,v,t)}{\partial q_a} \quad \Rightarrow \quad \sum_{b=1}^{f}\left(\frac{\partial^2 L}{\partial v_b \partial v_a}\frac{dv_b}{dt} + \frac{\partial^2 L}{\partial q_b \partial v_a}v_b\right) + \frac{\partial^2 L}{\partial t \partial v_a} = \frac{\partial L}{\partial q_a}$$

which, together with the equations

$$\frac{dq_a}{dt} = v_a,$$

constitute a set of 2f first order equations for the 2f variables $(q_a, v_a)$. These equations are not particularly attractive. The time derivatives of the v's in the first f "v-equations" are mixed up, and the other terms in these equations are messy. The form of Lagrange's equations suggests that more suitable choices for new variables are the **generalized momenta**

$$p_a = \frac{\partial L}{\partial \dot{q}_a}.$$

Recall that for a particle of mass m moving in a potential V and described by cartesian coordinates, the generalized momenta are the ordinary linear momenta. The definitions $p_a = \partial L/\partial \dot{q}_a$ can be inverted (provided $\det|\partial^2 L/\partial \dot{q}_b \partial \dot{q}_a| \neq 0$, which is usually the case[1]) to give

$$\dot{q}_a = \frac{dq_a}{dt} = F_a(q,p,t).$$

Lagrange's equations themselves can be written

$$\frac{dp_a}{dt} = \frac{\partial L}{\partial q_a} = G_a(q,p,t),$$

where in the second equality the $\dot{q}$'s in $\partial L/\partial q$ have been replaced by the F's. Together these form a set of 2f first order equations for the 2f variables $(q_a, p_a)$. In these equations the derivative of each variable in turn stands alone on the left. This in itself is not remarkable; the same could be accomplished for the set $(q_a, v_a)$ by multiplying the v-equations by the matrix reciprocal to $\partial^2 L/\partial v_b \partial v_a$. What is remarkable, as we now show, is the simple structure of the functions $F_a$ and $G_a$.

Introduce the **Hamiltonian**

$$H(q,p,t) = \sum_{a=1}^{f} p_a \dot{q}_a - L(q,\dot{q},t),$$

in which we have used $p_a = \partial L/\partial \dot{q}_a$ to express the $\dot{q}$'s on the right-hand side in terms of the q's and p's. The Hamiltonian H is thus a function of the q's and p's (and possibly also of t explicitly) and does not contain their time derivatives. From the definition of H we have

$$dH = \sum_{a=1}^{f} \left[ \dot{q}_a dp_a + \left( p_a - \frac{\partial L}{\partial \dot{q}_a} \right) d\dot{q}_a - \frac{\partial L}{\partial q_a} dq_a \right] - \frac{\partial L}{\partial t} dt.$$

In this equation we should really express $d\dot{q}$ in terms of the $dq$ and $dp$, but this turns out to be unnecessary since the coefficient of each of the $d\dot{q}_a$ is zero. We are left with

$$dH = \sum_{a=1}^{f} \left[ \dot{q}_a dp_a - \frac{\partial L}{\partial q_a} dq_a \right] - \frac{\partial L}{\partial t} dt.$$

On the other hand, since H is a function of q, p, and t, we have

---

[1]For what to do if this is not the case, see E. C. G. Sudarshan and N. Mukunda, *Classical Dynamics: A Modern Perspective* (John Wiley & Sons, New York, 1974), Chapter 8.

$$dH = \sum_{a=1}^{f}\left[\frac{\partial H}{\partial q_a}dq_a + \frac{\partial H}{\partial p_a}dp_a\right] + \frac{\partial H}{\partial t}dt.$$

Comparing the two expressions, we see that

$$\dot{q}_a = \frac{\partial H}{\partial p_a} \qquad \frac{\partial L}{\partial q_a} = -\frac{\partial H}{\partial q_a} \qquad \frac{\partial H}{\partial t} = -\frac{\partial L}{\partial t}.$$

Note that " $\partial/\partial q$ " in the second of these equations means different things on the two sides of the equation. On the left-hand side it means "differentiate with respect to q holding $\dot{q}$ fixed"; on the right it means "differentiate with respect to q holding p fixed." The Lagrangian L is to be regarded as a function of q, $\dot{q}$, and t; the Hamiltonian H is to be regarded as a function of q, p, and t. When we differentiate, we must hold the appropriate quantities fixed. The equations for the 2f variables $(q_a, p_a)$ can now be written

$$\frac{dq_a}{dt} = \frac{\partial H}{\partial p_a} \qquad \frac{dp_a}{dt} = -\frac{\partial H}{\partial q_a}.$$

This set of 2f first order equations for the 2f variables $(q_a, p_a)$, with its remarkable symmetry and structure, is known as **Hamilton's equations**. We see that the functions $F_a$ and $G_a$ can be obtained from a *single* function, the Hamiltonian H, by differentiation with respect to $p_a$ or $q_a$. Note also that the coordinates $q_a$ and momenta $p_a$ appear in Hamilton's equations on an essentially equal footing; the general form of Hamilton's equations remains unchanged if we replace $q_a$ by $p_a$ and $p_a$ by $-q_a$. This is an example of the transformations which we study in detail in Chapter VII.

Using Hamilton's equations we can easily show that

$$\frac{dH}{dt} = \frac{\partial H}{\partial t},$$

so that if the Hamiltonian does not depend explicitly on time, it is a constant of the motion. We have already noted this useful fact in Chapter V.

Hamilton's equations form the starting point for practically all investigations in advanced mechanics, indeed to such a degree that the equations are often called the **canonical equations** of motion. The set of 2f variables $(q_a, p_a)$ are then called the **canonical variables**; the generalized coordinates $q_a$ are the **canonical coordinates**; the generalized momenta $p_a$ are the **canonical momenta conjugate to** $q_a$.

The 2f dimensional space of the canonical variables $(q_a, p_a)$ is called **phase space**. The complete dynamical state of a system at any instant of time, say at some "initial" time $t_0$, is represented by a point in phase space. As time goes on, this point moves, tracing out a *unique* trajectory. There is thus just one trajectory through each point in phase space; the phase space trajectories do not intersect. The complete collection of phase space trajectories (or a representative number of them) provides a phase space portrait. Such portraits are very informative, particularly for one degree of freedom where they are easily drawn, and enable us to form an overall picture of the dynamical behavior of a system.

## Plane pendulum

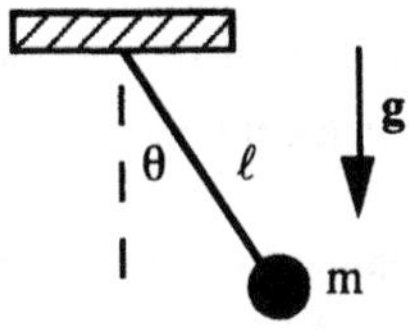


Fig. 6.01. Plane pendulum

Consider once more the plane pendulum (Fig. 6.01). The Lagrangian is

$$L = \tfrac{1}{2} m\ell^2\dot{\theta}^2 + mg\ell\cos\theta .$$

The momentum $p_\theta$ conjugate to $\theta$ is

$$p_\theta = \frac{\partial L}{\partial \dot{\theta}} = m\ell^2\dot{\theta}$$

and is physically the angular momentum about the point of support. The Hamiltonian is

$$H = p_\theta\dot{\theta} - L = \tfrac{1}{2} m\ell^2\dot{\theta}^2 - mg\ell\cos\theta$$

and is physically the total energy. Of course we must express H in terms of the proper variables $\theta$ and $p_\theta$. This is easily dore by using the definition of $p_\theta$ to express $\dot{\theta}$ in terms of $p_\theta$. We then get H in the required form,

$$H = \frac{p_\theta{}^2}{2m\ell^2} - mg\ell\cos\theta .$$

Hamilton's equations are

$$\frac{d\theta}{dt} = \frac{\partial H}{\partial p_\theta} = \frac{p_\theta}{m\ell^2} \qquad \frac{dp_\theta}{dt} = -\frac{\partial H}{\partial \theta} = -mg\ell\sin\theta .$$

Since the Hamiltonian does not depend explicitly on time, it is a constant of the motion. The trajectories of the system in the $(\theta, p_\theta)$ phase space are thus given by

$$\frac{p_\theta^2}{2m\ell^2} - mg\ell\cos\theta = E$$

where E, the energy, is the constant value of the Hamiltonian. These trajectories are shown in Fig. 6.02 for various values of the energy. Note that in Fig. 6.02(a) $\theta = \pi$ and $\theta = -\pi$ are the same physical point, so we should imagine Fig. 6.02(a) rolled into a cylinder and joined along the dashed lines. When a moving phase point "leaves" Fig. 6.02(a) at $(\theta = \pm\pi, p_\theta)$, it instantaneously "re-enters" at $(\theta = \mp\pi, p_\theta)$. Alternatively, we can imagine Fig. 6.02(a) repeated over and over in the $\theta$-direction, as in Fig. 6.02(b).

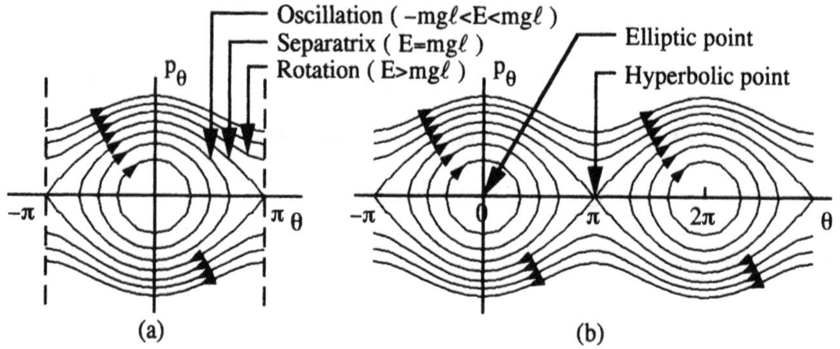


Fig. 6.02. Phase space for the plane pendulum

The minimum value of E is $-mg\ell$ and corresponds to the pendulum hanging vertically $(\theta = 0)$ at rest $(p_\theta = 0)$. For $E + mg\ell$ small compared to $mg\ell$, $\theta$ and $p_\theta$ stay close to $(0,0)$ and we can expand $\cos\theta \approx 1 - \theta^2/2$ to give

$$\frac{p_\theta^2}{2m\ell^2} + \frac{1}{2}mg\ell\theta^2 \approx E + mg\ell.$$

These low-energy phase space trajectories are ellipses around $(0,0)$ and correspond to the usual oscillatory motion of the pendulum with $\theta$ and $p_\theta$ out of phase by $\pi/2$. The point $(0,0)$ is sometimes called an **elliptic point**; it is a point of stable equilibrium. As the energy is increased, the amplitude of oscillation increases, until at $E = mg\ell$ the pendulum is barely able to reach $\theta = \pi$ (with $p_\theta = 0$). For $E > mg\ell$ the motion is no longer oscillatory but rotational, with the pendulum swinging round and round over the top. The phase space curves with $E = mg\ell$ which separate these two types of motion are called the **separatrices**. They would seem to violate (at the point $(\pm\pi, 0)$, seen most conveniently in Fig. 6.02(b)) the statement that only one phase space trajectory pass through any given point. If we look more closely at the region around $(\pm\pi, 0)$ (see Fig. 6.03(b)), however, we see that it consists of the following trajectories: the point itself, two "ends" of the separatrices which approach the point as a limit, but do not reach it in finite time, and two similar ends which recede from the point. Other nearby trajectories are hyperbolas, as can be seen by expanding the equation for the trajectories about the point $(\pm\pi, 0)$,[2]

$$\frac{p_\theta^2}{2m\ell^2} - \frac{1}{2} mg\ell(\pm\pi - \theta)^2 \approx E - mg\ell.$$

The point $(\pm\pi, 0)$ is sometimes called a **hyperbolic point**; it is an point of unstable equilibrium. The features of elliptic and hyperbolic equilibrium points are summarized in Fig. 6.03.

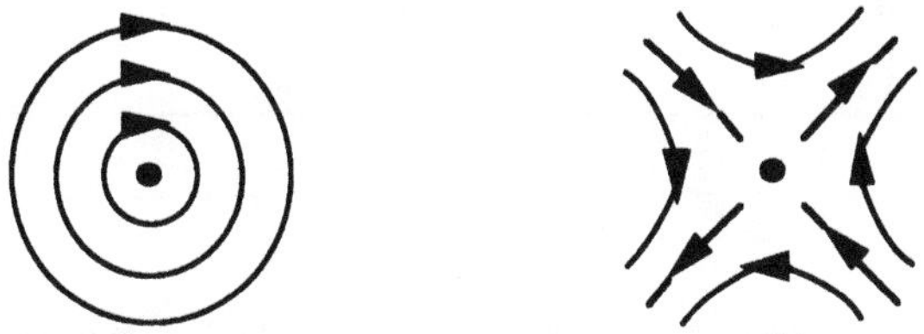
(a) (b)

Fig. 6.03. Phase space near (a) an elliptic point, (b) a hyperbolic point

## Spherical pendulum

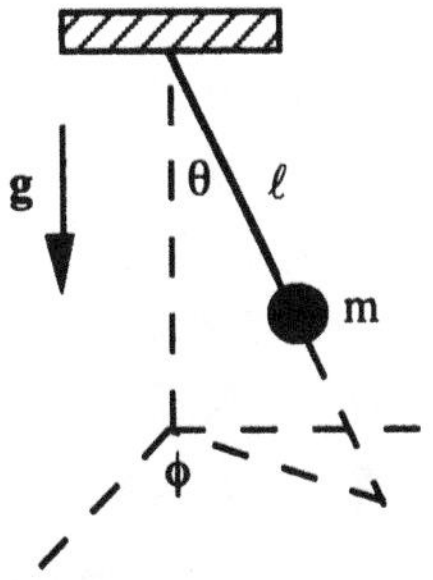


Fig. 6.04. Spherical pendulum

The Lagrangian for the spherical pendulum (Fig. 6.04) is

$$L = \tfrac{1}{2} m\ell^2 (\dot\theta^2 + \dot\phi^2 \sin^2\theta) + mg\ell\cos\theta.$$

The canonical momenta conjugate to $\theta$ and $\phi$ are

---

[2]Alternatively, one can obtain this by reversing gravity in the preceding equation.

$$p_\theta = \frac{\partial L}{\partial \dot\theta} = m\ell^2\dot\theta \qquad p_\phi = \frac{\partial L}{\partial \dot\phi} = m\ell^2 \sin^2\theta\,\dot\phi,$$

and the Hamiltonian is

$$H = \tfrac{1}{2}m\ell^2(\dot\theta^2 + \dot\phi^2 \sin^2\theta) - mg\ell\cos\theta = \frac{p_\theta{}^2}{2m\ell^2} + \frac{p_\phi{}^2}{2m\ell^2\sin^2\theta} - mg\ell\cos\theta .$$

Again, H is constant in time and equals the total energy. Hamilton's equations are

$$\frac{d\theta}{dt} = \frac{\partial H}{\partial p_\theta} = \frac{p_\theta}{m\ell^2} \qquad \frac{d\phi}{dt} = \frac{\partial H}{\partial p_\phi} = \frac{p_\phi}{m\ell^2\sin^2\theta}$$

$$\frac{dp_\theta}{dt} = -\frac{\partial H}{\partial \theta} = \frac{p_\phi{}^2\cos\theta}{m\ell^2\sin^2\theta} - mg\ell\sin\theta \qquad \frac{dp_\phi}{dt} = -\frac{\partial H}{\partial \phi} = 0 .$$

Note that $\phi$ is a cyclic coordinate. It does not appear in the Lagrangian, and hence it does not appear in the Hamiltonian either. The momentum $p_\phi$ conjugate to $\phi$ is a constant of the motion. Thus, in integrating the equations for the non-cyclic coordinate $\theta$ and momentum $p_\theta$, we can "ignore" the cyclic variable; the coordinate itself does not appear in the equations, and its conjugate momentum can be regarded as a constant parameter. For this reason cyclic coordinates are sometimes called **ignorable** coordinates.

## Rotating pendulum

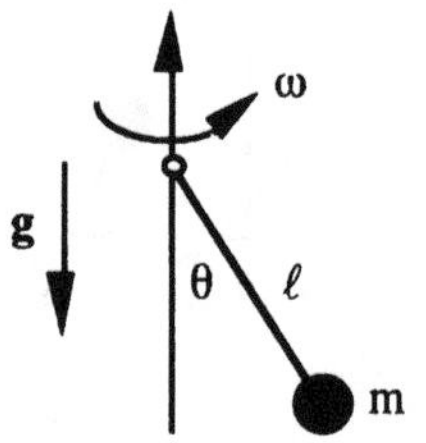


Fig. 6.05 Rotating pendulum

Now consider a pendulum which is constrained to swing in a plane which rotates about a vertical axis through the point of support at an angular rate $\omega$ as determined by some external drive mechanism. Choosing for the (single) generalized coordinate the angle $\theta$, we have the Lagrangian

$$L = \tfrac{1}{2}m\ell^2(\dot\theta^2 + \omega^2\sin^2\theta) + mg\ell\cos\theta .$$

The momentum conjugate to $\theta$ is again $p_\theta = m\ell^2\dot{\theta}$, and the Hamiltonian is

$$H = \frac{p_\theta^{\,2}}{2m\ell^2} - \frac{1}{2}m\ell^2\omega^2\sin^2\theta - mg\ell\cos\theta .$$

In this case the Hamiltonian is *not* equal to the total energy.[3] However, the Hamiltonian *is* constant in time, if the drive mechanism keeps the rotation rate $\omega$ constant. The system energy is then not constant. It is clear why: the drive mechanism can do work on the system. Indeed, this work can be written

$$\delta W = N\,d\phi = \frac{dJ}{dt}d\phi = \omega\,dJ ,$$

where N is the torque exerted by the drive mechanism on the system, and J is the angular momentum of the system about the vertical axis. Energy balance thus gives

$$dE = \omega\,dJ ,$$

which can be rewritten in the form

$$d(E - \omega J) = -J\,d\omega .$$

The quantity $E - \omega J$, which is like a "free energy" in thermodynamics, is thus constant if the rotation rate $\omega$ is held constant. It equals the Hamiltonian H, since the angular momentum of the system about the vertical axis is $J = m\ell^2\omega\sin^2\theta$.

It is worthwhile to pursue this problem a little further. The phase space trajectories are given by $H = \text{constant}$. In particular, the equilibrium points (the points where $\dot{\theta} = 0$ and $\dot{p}_\theta = 0$) satisfy

$$\frac{\partial H}{\partial p_\theta} = \frac{p_\theta}{m\ell^2} = 0$$

$$\frac{\partial H}{\partial \theta} = mg\ell\sin\theta\left[1 - (\omega^2\ell/g)\cos\theta\right] = 0 .$$

For small rotation rate $\omega$ these equations have only the solutions $(\theta, p_\theta) = (0,0)$ and $(\pm\pi, 0)$, as for the simple pendulum. Near the equilibrium point $(0,0)$ the Hamiltonian is given by

$$H + mg\ell \approx \frac{p_\theta^2}{2m\ell^2} + \frac{1}{2}mg\ell\left[1 - (\omega^2\ell/g)\right]\theta^2 .$$

[3]The sign of the middle term would be plus for energy.

For small $\omega$ the coefficient of $\theta^2$ is positive, and the phase space trajectories near (0,0) are ellipses around the point; the equilibrium point (0,0) is a stable elliptic point (Fig. 6.06(a)). As $\omega$ is increased, the coefficient of $\theta^2$ decreases, and at $\omega = \sqrt{g/\ell}$ (the angular frequency of a simple pendulum of the same length) it changes from positive to negative. The phase space trajectories near (0,0) change from ellipses to hyperbolas, and the point (0,0) changes from a stable elliptic to an unstable hyperbolic equilibrium point (Fig. 6.06(b)).

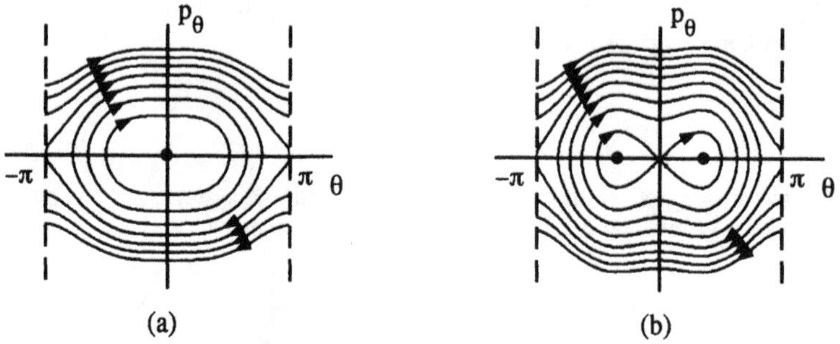


Fig. 6.06. Phase space for rotating pendulum
(a) for $\omega < \sqrt{g/\ell}$, (b) for $\omega > \sqrt{g/\ell}$

At the rotation rate $\omega = \sqrt{g/\ell}$ the equilibrium point (0,0) splits, and two new equilibrium points at $(\pm\theta_0, 0)$ with $\cos\theta_0 = g/\ell\omega^2$ come into being. We say that the equilibrium point (0,0) undergoes **bifurcation**. These new equilibrium points may be shown to be stable. The equilibrium point $(\pm\pi, 0)$ may be shown to be unstable at all rotation rates. The locations of the various equilibrium points are shown in Fig. 6.07 as functions of the rotation rate, with solid lines indicating stable points and dashed lines indicating unstable points.

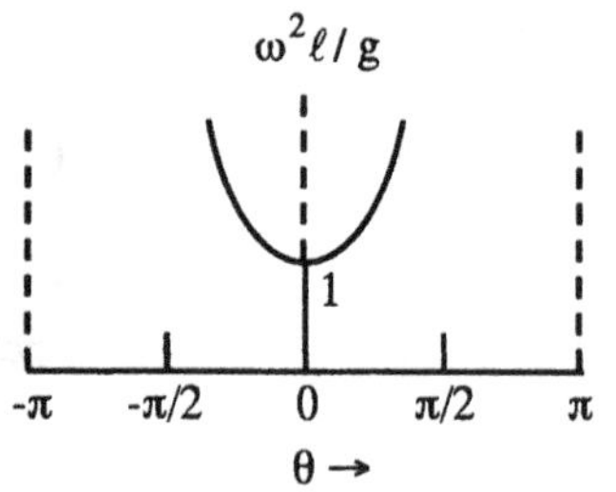


Fig. 6.07. Equilibrium points for rotating pendulum

Independent of the rotation rate, the Hamiltonian is invariant under spatial reflection, $\theta \rightarrow \theta' = -\theta$, $p_\theta \rightarrow p'_\theta = -p_\theta$, and for $\omega < \sqrt{g/\ell}$ the equilibrium points in phase space share this symmetry. However, for $\omega > \sqrt{g/\ell}$ this is no longer so. The equilibrium point $(\theta_0, 0)$ is not invariant under spatial reflection, but is transformed into *another* equilibrium point $(-\theta_0, 0)$. This is an example of **symmetry breaking**.

## Electromagnetic interaction

The Lagrangian for a particle of mass m and charge e in an electromagnetic field is

$$L = \tfrac{1}{2}mv^2 - e\phi + (e/c)\mathbf{A}\cdot\mathbf{v}$$

where $\phi$ and $\mathbf{A}$ are the scalar and vector potentials for the field. The momentum conjugate to $\mathbf{r}$ is

$$\mathbf{p} = \frac{\partial L}{\partial \mathbf{v}} = m\mathbf{v} + (e/c)\mathbf{A}$$

and is the sum of the kinetic momentum $m\mathbf{v}$ and the field momentum $(e/c)\mathbf{A}$, as discussed more fully in Chapter III. The Hamiltonian is then

$$H = \mathbf{p}\cdot\mathbf{v} - L = \tfrac{1}{2}mv^2 + e\phi = \frac{|\mathbf{p} - (e/c)\mathbf{A}|^2}{2m} + e\phi.$$

Note that this Hamiltonian can be obtained from the free particle Hamiltonian $H = |\mathbf{p}|^2/2m$ by making the replacements $H \to H - e\phi$, $\mathbf{p} \to \mathbf{p} - (e/c)\mathbf{A}$. This method of introducing the electromagnetic interaction is used frequently in quantum theory and leads to what is called **minimal electromagnetic interaction**. Note that the procedure (though not the above Hamiltonian) is relativistically invariant, $(H, c\mathbf{p})$ and $(\phi, \mathbf{A})$ both forming relativistic four-vectors.

## Poisson brackets

The time rate of change of any function $u(q,p)$ of the canonical variables is

$$\frac{du}{dt} = \sum_{a=1}^{f}\left(\frac{\partial u}{\partial q_a}\frac{dq_a}{dt} + \frac{\partial u}{\partial p_a}\frac{dp_a}{dt}\right) = \sum_{a=1}^{f}\left(\frac{\partial u}{\partial q_a}\frac{\partial H}{\partial p_a} - \frac{\partial u}{\partial p_a}\frac{\partial H}{\partial q_a}\right).$$

Mathematical structures like the right-hand side occur frequently in Hamiltonian mechanics and, as we shall see, have interesting properties. It is useful to have a special name and notation for them. We define the **Poisson bracket** $[u,v]$ of two functions $u(q,p)$ and $v(q,p)$ of the canonical variables by

$$[u,v] = \sum_{a=1}^{f}\left(\frac{\partial u}{\partial q_a}\frac{\partial v}{\partial p_a} - \frac{\partial u}{\partial p_a}\frac{\partial v}{\partial q_a}\right).$$

We can then write the above equation in the form

$$\frac{du}{dt} = [u,H].$$

In particular, Hamilton's equations of motion in Poisson bracket notation become

$$\frac{dq_a}{dt} = [q_a,H] \qquad \frac{dp_a}{dt} = [p_a,H].$$

Poisson brackets have the following properties:

1. Linearity:
   $[u_1 + u_2, v] = [u_1, v] + [u_2, v]$
   $[cu, v] = c[u, v]$ where c is a constant

2. Antisymmetry:
   $[u, v] = -[v, u]$

3. Product rule:
   $[u, vw] = [u, v]w + v[u, w]$

4. "Jacobi identity":
   $[u,[v,w]] + [v,[w,u]] + [w,[u,v]] = 0$

Relations 1, 2, and 3 follow readily from the definition of the Poisson bracket. Relation 4, the **Jacobi identity**, is less obvious. It can, however, be established in a straightforward way by expanding the double Poisson brackets and comparing terms.[4]

Those familiar with quantum mechanics will recognize all the above equations (except for the original definition of the Poisson bracket) as correct quantum mechanical equations. All that must be done is to interpret $[u, v]$ not as a classical Poisson bracket, but as a **commutator bracket**

$$[u, v]_{\text{Poisson}} \to (1/i\hbar)(uv - vu) = (1/i\hbar)[u, v]_{\text{commutator}}$$

of the non-commuting quantum mechanical operators u and v. This forms the basis of the Dirac-Heisenberg approach to quantum mechanics[5] and provides a way of transcribing classical mechanics into quantum mechanics,[6] when a classical limit exists.

[4]A more sophisticated (though no shorter) proof can be found in L. D. Landau and E. M. Lifshitz, *Mechanics*, (Pergamon Press, Oxford, 1960; 1969; 1976), 3rd ed., trans. J. B. Sykes and J. S. Bell., p. 136.

[5]P. A. M. Dirac, *The Principles of Quantum Mechanics*, (Oxford University Press, London, 1930; 1935; 1947; 1958), 4th ed., sect. 21; also see the papers by W. Heisenberg; M. Born and P. Jordan; P. A. M. Dirac; and M. Born, W. Heisenberg and P. Jordan, reprinted in *Sources of Quantum Mechanics*, (North-

These general properties, together with the easily obtained **fundamental Poisson brackets**

$$[q_a,q_b]=0 \qquad [q_a,p_b]=\delta_{ab} \qquad [p_a,p_b]=0,$$

can often be used to evaluate Poisson brackets without referring back to the original definition. The advantage of this approach is that it applies to both the classical and quantum situations. For example, the Poisson bracket of the x-component $L_x$ and the y-component $L_y$ of the angular momentum of a particle is

$$\begin{aligned}[L_x,L_y] &= [yp_z - zp_y, zp_x - xp_z] \\ &= [yp_z,zp_x]-[yp_z,xp_z]-[zp_y,zp_x]+[zp_y,xp_z] && \text{(Properties 1 and 2)} \\ &= 16 \text{ terms, the only non-vanishing ones} && \text{(Property 3)} \\ &\quad \text{being } y[p_z,z]p_x + x[z,p_z]p_y \\ &= xp_y - yp_x && \text{(Fundamental bracket)} \\ &= L_z \end{aligned}$$

with two similar relations obtained by cyclic permutation of x,y,z.

The Jacobi identity leads to the interesting theorem: if $u(q,p)$ and $v(q,p)$ are constants of the motion, then their Poisson bracket $[u,v]$ is also a constant of the motion. The proof is simple: if u and v are constants of the motion, their Poisson brackets with the Hamiltonian vanish. The Jacobi identity,

$$[u,[v,H]]+[v,[H,u]]+[H,[u,v]]=0,$$

then shows that the Poisson bracket of $[u,v]$ with the Hamiltonian vanishes, so that $[u,v]$ is a constant of the motion as well. For example, if $L_x$ and $L_y$ are constants of the motion, then so is $L_z$. As in this case, the theorem sometimes enables us to find new constants of the motion. Quite often, however, it simply yields zero or constants of the motion we already know.

---

Holland Publishing Co., Amsterdam, 1967; republished by Dover Publications, New York, NY, 1968), ed. B. L. Van der Waerden.

[6] The order of the factors on the right hand side of Relation 3, w after and v before, makes no difference for Poisson brackets, but has been chosen so that the relation is also correct for commutator brackets.

## Exercises

1. A system with one degree of freedom has a Hamiltonian
$$H(q,p) = \frac{p^2}{2m} + A(q)p + B(q)$$
where A and B are certain functions of the coordinate q, and p is the momentum conjugate to q.
(a) Find the velocity $\dot{q}$.
(b) Find the Lagrangian $L(q,\dot{q})$ (note variables).

2. We have seen (exercise 5.01) that two Lagrangians $L'$ and L which differ by the total time derivative $d\Lambda/dt$ of some function $\Lambda(q,t)$,
$$L' = L + d\Lambda/dt,$$
are equivalent, leading to the same Lagrange's equations of motion.
(a) What is the relation between the generalized momenta $p'$ and p which these two Lagrangians yield?
(b) What is the relation between the Hamiltonians $H'$ and H which these two Lagrangians yield?
(c) Show explicitly that Hamilton's equations of motion in the primed quantities are equivalent to those in the unprimed quantities.

3. A particle of mass m moves in a central force field with potential V(r). The Lagrangian in terms of spherical polar coordinates $(r,\theta,\phi)$ is
$$L = \tfrac{1}{2}m\left(\dot{r}^2 + r^2\dot{\theta}^2 + r^2\sin^2\theta\dot{\phi}^2\right) - V(r).$$
(a) Find the momenta $(p_r, p_\theta, p_\phi)$ conjugate to $(r,\theta,\phi)$.
(b) Find the Hamiltonian $H(r,\theta,\phi,p_r,p_\theta,p_\phi)$.
(c) Write down the explicit Hamilton's equations of motion.

4. The Lagrangian for a free particle in terms of paraboloidal coordinates $(\xi,\eta,\phi)$ is (see exercise 3.09)
$$L = \tfrac{1}{2}m\left(\xi^2 + \eta^2\right)\left(\dot{\xi}^2 + \dot{\eta}^2\right) + \tfrac{1}{2}m\xi^2\eta^2\dot{\phi}^2.$$
(a) Find the momenta conjugate to $(\xi,\eta,\phi)$.
(b) Find the Hamiltonian.

5. The Lagrangian for a free particle of mass m, referred to cartesian coordinates $(x',y',z')$ which are rotating about an inertial z-axis with angular velocity $\omega$, is (see exercise 3.13)
$$L = \tfrac{1}{2}m[(\dot{x}'^2 + \dot{y}'^2 + \dot{z}'^2) + 2\omega(x'\dot{y}' - y'\dot{x}') + \omega^2(x'^2 + y'^2)].$$
(a) Find the momenta $(p'_x, p'_y, p'_z)$ conjugate to $(x',y',z')$.
(b) Find the Hamiltonian $H(x',y',z',p'_x,p'_y,p'_z)$.
(Ans. $H = \frac{1}{2m}\left(p_x'^2 + p_y'^2 + p_z'^2\right) - \omega\left(x'p'_y - y'p'_x\right)$)

6. The equations of motion for a particle of mass m and charge e moving in a uniform magnetic field B which points in the z-direction can be obtained from a Lagrangian (see exercise 3.14)

$$L = \tfrac{1}{2}m(\dot{x}^2 + \dot{y}^2 + \dot{z}^2) + (eB/2c)(x\dot{y} - y\dot{x}).$$

(a) Find the momenta $(p_x, p_y, p_z)$ conjugate to $(x, y, z)$.

(b) Find the Hamiltonian, expressing your answer first in terms of $(x, y, z, \dot{x}, \dot{y}, \dot{z})$ and then in terms of $(x, y, z, p_x, p_y, p_z)$.

(c) Evaluate the Poisson brackets

(i) $[m\dot{x}, m\dot{y}]$

(ii) $[m\dot{x}, H]$

7. Consider the one-dimensional motion of a particle in the following potential wells, in each case sketching representative trajectories in $(x, p)$ phase space:

(a) an infinite square well $V(x) = 0$ for $0 < x < a$
$V(x) \to \infty$ for $x \le 0$ and for $x > a$

(b) a bouncing ball $V(x) = mgx$ for $x > 0$
$V(x) \to \infty$ for $x \le 0$

(c) a simple harmonic oscillator $V(x) = \tfrac{1}{2}kx^2$

(d) a double well $V(x) = -\tfrac{1}{2}kx^2 + \tfrac{1}{4}k\dfrac{x^4}{a^2}$

8. In the potential wells of exercise 6.07 the motion is periodic but not necessarily simple harmonic. The **action variable** I is defined by

$$I = \frac{1}{2\pi}\oint p\,dx$$

where p is the momentum, and the integration is over a single period of the motion (see Chapter IX for further details).

(a) Show that the action variable is the *area enclosed* by the orbit in phase space divided by $2\pi$ and is given by

$$I = \frac{1}{\pi}\int_{x_1(E)}^{x_2(E)} \sqrt{2m(E - V(x))}\,dx,$$

where E is the total energy, $V(x)$ is the potential energy, and $x_1(E)$ and $x_2(E)$ are the lower and upper turning points of the motion.

(b) Show that the period $\tau$ of the motion is given by

$$\tau = 2\pi\frac{dI}{dE}.$$

9. (a) Evaluate the action variable $I(E)$ (see exercise 6.08) for:

(i) an infinite square well $V(x) = 0$ for $0 < x < a$
$V(x) \to \infty$ for $x \le 0$ and for $x > a$

(ii) a bouncing ball $V(x) = mgx$ for $x > 0$
$V(x) \to \infty$ for $x \le 0$

(iii) a simple harmonic oscillator $V(x) = \frac{1}{2}kx^2$
and use your results to find the periods of the motions.
(b) In the "old quantum mechanics" of Bohr and Sommerfeld the action variable I was quantized in units of $\hbar$ (Planck's constant divided by $2\pi$). What does this give for the energy levels of the systems of part (a)?

10. (a) Use the definition of the Poisson bracket to establish Poisson bracket properties 1, 2, 3, and also the Jacobi identity 4.
(b) Show that these four properties also hold for the commutator bracket.

11. Let $f(q(t),p(t))$ be some function of the canonical variables, and $f_0$ its value at time 0.
(a) Show, if the Hamiltonian H is time-independent, that the function f at time t is given by

$$f = f_0 + t[f_0,H] + (t^2/2!)[[f_0,H],H] + (t^3/3!)[[[f_0,H],H],H] + \cdots.$$

(b) A particle of mass m moving in one dimension x is acted on by a constant force F. The Hamiltonian is $H = p^2/2m - Fx$. Suppose that at time 0 the particle is at $x_0$ with momentum $p_0$. Use the result of (a) to find the position x and momentum p at time t.
(c) A particle of mass m moving in one dimension x is in a simple harmonic oscillator well. The Hamiltonian is $H = p^2/2m + m\omega^2x^2/2$. Suppose that at time 0 the particle is at $x_0$ with momentum $p_0$. Use the result of (a) to find the position x and momentum p at time t.

12. The Hamiltonian for a simple harmonic oscillator is

$$H = \frac{p^2}{2m} + \frac{1}{2}m\omega^2x^2.$$

Introduce the complex quantities

$$a = \sqrt{\frac{m\omega}{2}}\left(x + \frac{ip}{m\omega}\right) \quad \text{and} \quad a^* = \sqrt{\frac{m\omega}{2}}\left(x - \frac{ip}{m\omega}\right).$$

(a) Express H in terms of a and $a^*$.
(b) Evaluate the Poisson brackets $[a,a^*]$, $[a,H]$, and $[a^*,H]$.
(c) Write down and solve the equations of motion for a and $a^*$.

13. (a) Evaluate the set of Poisson brackets for a component of the radius vector $\mathbf{r} = (x,y,z)$ with a component of the angular momentum $\mathbf{L} = (L_x,L_y,L_z)$. Also evaluate those for a component of the linear momentum $\mathbf{p} = (p_x,p_y,p_z)$ with a component of the angular momentum. Show that the results can be put in the form

$$[\mathbf{r},\mathbf{L}\cdot\mathbf{n}] = \mathbf{n}\times\mathbf{r} \qquad [\mathbf{p},\mathbf{L}\cdot\mathbf{n}] = \mathbf{n}\times\mathbf{p}$$

where $\mathbf{n}$ is an arbitrary constant vector.
(b) Use the results of (a) to show that the Poisson bracket of a component of the angular momentum with an arbitrary scalar function of $\mathbf{r}$ and $\mathbf{p}$, of the form $a(r^2,\mathbf{r}\cdot\mathbf{p},p^2)$, is zero.

(c) Use the results of (a) to show that the Poisson bracket of a component of the angular momentum with an arbitrary vector function of **r** and **p**, of the form

$$\mathbf{A} = a_1\mathbf{r} + a_2\mathbf{p} + a_3\mathbf{r} \times \mathbf{p},$$

is given by

$$[\mathbf{A}, \mathbf{L}\cdot\mathbf{n}] = \mathbf{n} \times \mathbf{A}.$$

(d) Show that the Poisson bracket of the square of the angular momentum $L^2 = L_x^2 + L_y^2 + L_z^2$ with an arbitrary vector function **A** of **r** and **p** is given by

$$[\mathbf{A}, L^2] = 2\mathbf{L} \times \mathbf{A}.$$

14. Consider motion of a particle of mass m in an isotropic harmonic oscillator potential $V = \frac{1}{2}kr^2$ and take the orbital plane to be the x-y plane. The Hamiltonian is then

$$H = S_0 = \frac{1}{2m}(p_x^2 + p_y^2) + \frac{1}{2}k(x^2 + y^2).$$

Introduce the three quantities

$$S_1 = \frac{1}{2m}(p_x^2 - p_y^2) + \frac{1}{2}k(x^2 - y^2), \quad S_2 = \frac{1}{m}p_x p_y + kxy, \quad S_3 = \omega(xp_y - yp_x)$$

with $\omega = \sqrt{k/m}$.

(a) Show that

$$[S_0, S_i] = 0 \qquad i = 1,2,3$$

so $(S_1, S_2, S_3)$ are constants of the motion.

(b) Show that

$$[S_1, S_2] = 2\omega S_3 \qquad [S_2, S_3] = 2\omega S_1 \qquad [S_3, S_1] = 2\omega S_2$$

so $(2\omega)^{-1}(S_1, S_2, S_3)$ have the same Poisson bracket relations as the components of a "three dimensional angular momentum."

(c) Show that

$$S_0^2 = S_1^2 + S_2^2 + 3_3^2.$$

(The corresponding quantum relation has $(\hbar\omega)^2$ added to the right-hand side.)

15. Consider motion of a particle of mass m in a gravitational potential $V = -k/r$ and take the orbital plane to be the x-y plane. The Hamiltonian is then

$$H = \frac{1}{2m}(p_x^2 + p_y^2) - \frac{k}{r}$$

where now $r = \sqrt{x^2 + y^2}$. The angular momentum vector points in the z-direction and has (z-) component

$$L = xp_y - yp_x,$$

and the Laplace-Runge-Lenz vector (see exercise 1.12) lies in the x-y plane and has components

$$K_x = p_y L - mk\,x/r, \qquad K_y = -p_x L - mk\,y/r.$$

(a) Show that

$$[L, H] = 0, \qquad [K_x, H] = 0, \qquad [K_y, H] = 0,$$

so L, $K_x$, and $K_y$ are constants of the motion.

(b) Show that

$$[K_x,L] = -K_y, \quad [K_y,L] = K_x, \quad [K_x,K_y] = -(2mH)L$$

(see exercise 6.13 for some useful Poisson brackets). Now restrict yourself to bound states of energy $H = -|E|$ and show that $\left(\frac{K_x}{\sqrt{2m|E|}}, \frac{K_y}{\sqrt{2m|E|}}, L\right)$ have the same Poisson bracket relations as the components of a "three dimensional angular momentum."

(c) Show that the square of the length of this "three dimensional angular momentum" is $\frac{mk^2}{2|E|}$. (The corresponding quantum relation is $\frac{mk^2}{2|E|} - \frac{1}{4}\hbar^2$.)

Exercises 6-14 and 6-15 can be extended to three dimensions. See, for example, Leonard I. Schiff, *Quantum Mechanics*, (McGraw-Hill Book Company, New York, 1968), 3rd ed., pp. 234-242.

# CHAPTER VII

# CANONICAL TRANSFORMATIONS

In Lagrangian dynamics we are free to choose generalized coordinates essentially as we wish. If q is a set of generalized coordinates, then any reversible point transformation $Q = Q(q,t)$ gives another set. Under such transformations Lagrange's equations of motion maintain their general form with Lagrangians related by $\bar{L}(Q,\dot{Q},t) = L(q,\dot{q},t)$. Similar freedom also exists in Hamiltonian dynamics. Indeed, since there are twice as many canonical variables $(q,p)$ as generalized coordinates, the set of possible transformations is considerably larger. This is one of the advantages of the canonical formalism. In this chapter we study these canonical transformations in detail. One result, more fully realized in Chapters VIII and IX, is the possibility of transforming to variables for which the equations of motion take on a simple form, thereby making possible the solution to the dynamical problem.

## One degree of freedom

There are many approaches to canonical transformations. Before studying them in all their generality, it is worthwhile to get an overview by restricting ourselves to systems with one degree of freedom and to transformations which do not involve the time explicitly. That is, we consider a transformation from old variables $(q,p)$ to new variables $(Q,P)$, of the form

$$Q = Q(q,p) \qquad P = P(q,p),$$

and ask what restrictions we must impose in order that this be a canonical transformation, that the new variables form a canonical set if the old ones do. The new variables must satisfy dynamical equations of canonical form, Hamilton's equations, so we begin by looking at the time rates of change of Q and P

$$\frac{dQ}{dt} = [Q,H]_{q,p} = \frac{\partial Q}{\partial q}\frac{\partial H}{\partial p} - \frac{\partial Q}{\partial p}\frac{\partial H}{\partial q}$$

$$\frac{dP}{dt} = [P,H]_{q,p} = \frac{\partial P}{\partial q}\frac{\partial H}{\partial p} - \frac{\partial P}{\partial p}\frac{\partial H}{\partial q} \,.$$

The partial derivatives of the Hamiltonian can be rewritten

$$\frac{\partial H}{\partial q} = \frac{\partial \bar{H}}{\partial Q}\frac{\partial Q}{\partial q} + \frac{\partial \bar{H}}{\partial P}\frac{\partial P}{\partial q}$$

$$\frac{\partial H}{\partial p} = \frac{\partial \bar{H}}{\partial Q}\frac{\partial Q}{\partial p} + \frac{\partial \bar{H}}{\partial P}\frac{\partial P}{\partial p}$$

where $\overline{H}(Q,P) = H(q(Q,P), p(Q,P))$ is the original Hamiltonian expressed in terms of the new variables. Substituting these expressions into the above equations gives

$$\frac{dQ}{dt} = \frac{\partial\overline{H}}{\partial P}\left(\frac{\partial Q}{\partial q}\frac{\partial P}{\partial p} - \frac{\partial Q}{\partial p}\frac{\partial P}{\partial q}\right) = \frac{\partial\overline{H}}{\partial P}[Q,P]_{q,p}$$

$$\frac{dP}{dt} = -\frac{\partial\overline{H}}{\partial Q}\left(\frac{\partial Q}{\partial q}\frac{\partial P}{\partial p} - \frac{\partial Q}{\partial p}\frac{\partial P}{\partial q}\right) = -\frac{\partial\overline{H}}{\partial Q}[Q,P]_{q,p}$$

where $[Q,P]_{q,p}$ is the Poisson bracket of the new variables with respect to the old. The new variables thus satisfy Hamilton's canonical equations of motion *independent of the form of the Hamiltonian* if, and only if, $[Q,P]_{q,p} = k$ where k is any non-zero constant.[1] Usually we are not interested in the changes in scale associated with $k \neq 1$, and carrying along an arbitrary constant is a nuisance. We thus take $k = 1$, *defining* a **canonical transformation** in one degree of freedom as a transformation $(q,p) \rightarrow (Q,P)$ for which

$$[Q,P]_{q,p} = 1.$$

The canonical equations of motion are then covariant under canonical transformations together with appropriate scale transformations.

Another way to approach canonical transformations is to observe that

$$\frac{\partial Q}{\partial q}\frac{\partial P}{\partial p} - \frac{\partial Q}{\partial p}\frac{\partial P}{\partial q} = 1$$

is the condition for the differential form

$$p\,dq - P\,dQ = \left(p - P\frac{\partial Q}{\partial q}\right)dq - P\frac{\partial Q}{\partial p}dp$$

---

[1]The "if" is trivial. To see the "only if," note that for the right hand sides to have the form $\partial K/\partial P$, $-\partial K/\partial Q$ (i.e. in order for a new Hamiltonian $K(Q,P)$ to exist), the cross-derivatives must be equal,

$$\frac{\partial}{\partial Q}\left(\frac{\partial\overline{H}}{\partial P}[Q,P]_{q,p}\right) = \frac{\partial}{\partial P}\left(\frac{\partial\overline{H}}{\partial Q}[Q,P]_{q,p}\right).$$

Expanding this gives

$$\frac{\partial\overline{H}}{\partial P}\frac{\partial[Q,P]_{q,p}}{\partial Q} = \frac{\partial\overline{H}}{\partial Q}\frac{\partial[Q,P]_{q,p}}{\partial P}.$$

We want this to be true independent of the form of $\overline{H}$ and hence require

$$\frac{\partial[Q,P]_{q,p}}{\partial Q} = 0 \qquad \frac{\partial[Q,P]_{q,p}}{\partial P} = 0,$$

so $[Q,P]_{q,p}$ must be independent of Q and P; that is, it is a constant.

to be an exact differential. Recall that the usual statement of the condition is that the cross-derivatives be equal,

$$\frac{\partial}{\partial p}\left(p - P\frac{\partial Q}{\partial q}\right) = \frac{\partial}{\partial q}\left(-P\frac{\partial Q}{\partial p}\right),$$

which is another way of writing the Poisson bracket condition. We can thus set

$$p\,dq - P\,dQ = dF$$

where F is a so-called **generating function** of the transformation. The reason for this name is as follows. Suppose that F is a function of the mixed set of variables $(q,Q)$. The right-hand side of the preceding equation then becomes

$$dF = \frac{\partial F}{\partial q}dq + \frac{\partial F}{\partial Q}dQ,$$

and we have

$$p = \frac{\partial F(q,Q)}{\partial q} \qquad P = -\frac{\partial F(q,Q)}{\partial Q}.$$

The first of these can be inverted to give $Q(q,p)$ (provided $\partial^2 F/\partial q\,\partial Q \neq 0$), and the second then gives $P(q,p)$. So in this sense F "generates the transformation."

As a specific example consider the transformation

$$q = \sqrt{\frac{2P}{m\omega}}\sin Q \qquad p = \sqrt{2m\omega P}\cos Q,$$

where it is more convenient to give the old variables $(q,p)$ in terms of the new variables $(Q,P)$. The constant $m\omega$ is arbitrary but this notation turns out to be useful for what follows. We evaluate the Poisson bracket

$$\begin{aligned}[q,p]_{Q,P} &= \frac{\partial q}{\partial Q}\frac{\partial p}{\partial P} - \frac{\partial q}{\partial P}\frac{\partial p}{\partial Q} \\ &= \sqrt{\frac{2P}{m\omega}}\cos Q \times \sqrt{\frac{m\omega}{2P}}\cos Q - \frac{1}{\sqrt{2m\omega P}}\sin Q \times \sqrt{2m\omega P}(-\sin Q) \\ &= 1\,.\end{aligned}$$

Thus the transformation (from new to old and hence also from old to new) is canonical. That this is so can also be seen by finding a generating function. Let us choose q and Q as independent variables, setting

$$P = \tfrac{1}{2} m\omega q^2 \csc^2 Q \qquad p = m\omega q \cot Q.$$

The differential form becomes

$$p\,dq - P\,dQ = m\omega q \cot Q\,dq - \tfrac{1}{2} m\omega q^2 \csc^2 Q\,dQ = d\left(\tfrac{1}{2} m\omega q^2 \cot Q\right).$$

It is an exact differential, so the transformation is canonical with generating function

$$F(q,Q) = \tfrac{1}{2} m\omega q^2 \cot Q.$$

The inverse transformation has the form

$$Q = \cot^{-1}\frac{p}{m\omega q} \qquad P = \frac{1}{\omega}\left(\frac{p^2}{2m} + \frac{1}{2} m\omega^2 q^2\right)$$

and suggests the following application. Suppose that $(q,p)$ are the usual canonical variables for a simple harmonic oscillator of mass m and angular frequency $\omega$. The Hamiltonian is

$$H = \frac{p^2}{2m} + \frac{1}{2} m\omega^2 q^2.$$

In terms of the new variables $(Q,P)$, the Hamiltonian becomes

$$\overline{H} = \omega P.$$

The new Hamiltonian $\overline{H}$ does not contain the new coordinate Q; the new coordinate is a cyclic coordinate. Hamilton's equations for the new variables are

$$\frac{dQ}{dt} = \frac{\partial \overline{H}}{\partial P} = \omega \qquad \frac{dP}{dt} = -\frac{\partial \overline{H}}{\partial Q} = 0$$

and can be integrated immediately to give

$$Q = \omega t + \beta \qquad P = \alpha$$

where $\alpha$ and $\beta$ are constants of integration. The equations of the canonical transformation then give the solution to the equations of motion for the original variables,

$$q = \sqrt{\frac{2\alpha}{m\omega}} \sin(\omega t + \beta) \qquad p = \sqrt{2m\omega\alpha}\,\cos(\omega t + \beta).$$

The new momentum P is an example of a canonical variable called the **action variable**. The coordinate Q conjugate to it is called the **angle variable**, and the above is a transformation to **action-angle variables**. This very useful transformation is discussed in detail in Chapter IX.

## Generating functions

We now turn to the general theory of canonical transformations. Consider a mechanical system with f degrees of freedom, which is described by a set of 2f canonical variables $(q_a, p_a)$. These variables satisfy Hamilton's canonical equations of motion

$$\frac{dq_a}{dt} = \frac{\partial H}{\partial p_a} \qquad \frac{dp_a}{dt} = -\frac{\partial H}{\partial q_a}$$

with Hamiltonian $H(q,p,t)$. Suppose we introduce a new set of variables $(Q_a, P_a)$ via the transformation

$$Q_a = Q_a(q,p,t) \qquad P_a = P_a(q,p,t).$$

We want the new variables, at the very least, to satisfy dynamical equations of canonical form

$$\frac{dQ_a}{dt} = \frac{\partial K}{\partial P_a} \qquad \frac{dP_a}{dt} = -\frac{\partial K}{\partial Q_a}$$

where K, the new Hamiltonian, is some function of Q, P, and t; that is, we want the new variables to be *canonical* variables. What are the restrictions on the transformation in order that this be so? Discovering these is made easier by the observation that Hamilton's equations can be regarded as Lagrange's equations for a system with 2f *independent* "generalized coordinates" $(q_a, p_a)$ and "Lagrangian"

$$L = \sum_{a=1}^{f} p_a \dot{q}_a - H(q,p,t).$$

We have

$$0 = \frac{d}{dt}\left(\frac{\partial L}{\partial \dot{q}_a}\right) - \frac{\partial L}{\partial q_a} = \frac{dp_a}{dt} - \left(-\frac{\partial H}{\partial q_a}\right)$$

$$0 = \frac{d}{dt}\left(\frac{\partial L}{\partial \dot{p}_a}\right) - \frac{\partial L}{\partial p_a} = 0 - \left(\dot{q}_a - \frac{\partial H}{\partial p_a}\right)$$

which are indeed Hamilton's equations. We can thus take over into Hamiltonian mechanics what we have already learned in Lagrangian mechanics[2] and, in particular, what we have learned about transformation of coordinates. We saw that in Lagrangian mechanics we could use essentially any set of generalized coordinates to describe a system; Lagrange's equations take the same general form no matter what set we use. Does this mean that there are *no* restrictions on canonical transformations? No, because Hamilton's equations arise from a Lagrangian with a very special form; if we make an arbitrary transformation of the canonical variables, the resulting Lagrangian

$$\bar{L}(Q,P,\dot{Q},\dot{P},t) = L(q,p,\dot{q},\dot{p},t)$$

does not usually have the required form. In order to yield equations of canonical form, the new Lagrangian $\bar{L}$ must be equivalent to

$$\sum_{a=1}^{f} P_a\dot{Q}_a - K(Q,P,t).$$

This does not mean that $\bar{L}$ must have exactly this form; $\bar{L}$ may differ from this by the total time derivative $dF/dt$ of some function F and still give the correct equations of motion. It is also possible to allow the two Lagrangians to differ by a constant factor. As pointed out earlier, however, such a factor is a nuisance to carry along, so we simply *define* a **canonical transformation** as a transformation $(q,p) \to (Q,P)$ for which

$$\sum_{a=1}^{f} p_a\dot{q}_a - H = \sum_{a=1}^{f} P_a\dot{Q}_a - K + \frac{dF}{dt}.$$

This can be written

$$\sum_{a=1}^{f} p_a\, dq_a - H\, dt = \sum_{a=1}^{f} P_a\, dQ_a - K\, dt + dF,$$

which says that a canonical transformation is a transformation which preserves the **(extended) canonical one-form** $\sum_{a=1}^{f} p_a\, dq_a - H\, dt$ up to a total differential. The function F is called the **generating function** of the transformation. Hamilton's equations are covariant under canonical transformations and also under more general transformations which include a change of scale.

---

[2]So, for example, Hamilton's equations can be obtained from a variational principle

$$\delta \int_{t_0}^{t_1} \left[ \sum_{a=1}^{f} p_a\dot{q}_a - H(q,p,t) \right] dt = 0$$

in which the q and p are regarded as a set of 2f independent variables, and the comparison paths are all those neighboring paths in (extended) phase space with the same end points as the path of interest.

The 4f variables (q,p) and (Q,P) are connected by the 2f equations of transformation. Thus, 2f of them are independent variables and the remaining 2f are dependent. *Which* 2f variables are independent and *which* are dependent depends partly on the particular transformation we are considering and partly on which variables we *choose* to regard as independent. For example, for the identity transformation $Q = q,\ P = p$ we cannot take $(q,Q)$ or $(p,P)$ as the independent variables, but $(q,p)$, or $(Q,P)$, or $(q,P)$, or $(p,Q)$ are all possible choices. The natural choices, always possible, would seem to be either the old variables $(q,p)$ or the new variables $(Q,P)$, and we eventually discuss this possibility. For our purposes here, however, a mixed set of variables, half old and half new, turns out to be more convenient.

Let us first suppose that the transformation is such that we can regard the old coordinates q and the new coordinates Q as independent. Such canonical transformations are often called **type 1**. For these the generating function $F_1$ is a function of q, Q, and t, so we can set in the preceding equation

$$dF_1(q,Q,t) = \sum_{a=1}^{f}\left(\frac{\partial F_1}{\partial q_a}dq_a + \frac{\partial F_1}{\partial Q_a}dQ_a\right) + \frac{\partial F_1}{\partial t}dt.$$

The coefficients of each of the $dq_a$, $dQ_a$, and $dt$ on the two sides of the equation must be equal. This gives

$$p_a = \frac{\partial F_1(q,Q,t)}{\partial q_a} \qquad P_a = -\frac{\partial F_1(q,Q,t)}{\partial Q_a} \qquad K = H + \frac{\partial F_1}{\partial t}.$$

The first set of f equations can be inverted (provided $\det|\partial^2 F_1/\partial q_a \partial Q_b| \neq 0$) to give the new coordinates Q in terms of the old coordinates q and momenta p; substituting this result into the second set of f equations then gives the new momenta P in terms of q and p; finally, the third equation gives the new Hamiltonian. Thus, $F_1$ "generates the transformation."

For the particular transformation we are considering, it may not be possible to regard (q,Q) as independent. For this or for other reasons (such as personal preference) we next consider transformations for which we can regard the old coordinates q and the new momenta P as independent. Such canonical transformations are often called **type 2**. It is convenient to rewrite the left-hand side of the equation which defines a canonical transformation by setting

$$\sum_{a=1}^{f} P_a\, dQ_a = d\left(\sum_{a=1}^{f} P_a Q_a\right) - \sum_{a=1}^{f} Q_a\, dP_a .$$

We then have

$$\sum_{a=1}^{f} p_a\, dq_a + \sum_{a=1}^{f} Q_a\, dP_a + (K - H)dt = dF_2$$

where

$$F_2 = F_1 + \sum_{a=1}^{f} P_a Q_a$$

is the generating function of the type 2 transformation. Note that it is *not* obtained simply by changing variables in the type 1 generating function. If we take $F_2$ to be a function of the appropriate independent variables $(q, P, t)$, we have

$$dF_2(q,P,t) = \sum_{a=1}^{f}\left(\frac{\partial F_2}{\partial q_a}dq_a + \frac{\partial F_2}{\partial P_a}dP_a\right) + \frac{\partial F_2}{\partial t}dt.$$

Comparing with the above, we then obtain the equations of the type 2 canonical transformation

$$p_a = \frac{\partial F_2(q,P,t)}{\partial q_a} \qquad Q_a = \frac{\partial F_2(q,P,t)}{\partial P_a} \qquad K = H + \frac{\partial F_2}{\partial t}.$$

The first set of f equations can be inverted (provided $\det|\partial^2 F_2/\partial q_a \partial P_b| \neq 0$) to give the new momenta P in terms of the old coordinates q and momenta p; the second set of f equations then gives the new coordinates Q in terms of q and p; and the third equation gives the new Hamiltonian.

It is clear that we could continue by considering transformations for which $(p, Q)$ are independent or for which $(p, P)$ are independent, or by considering more generally transformations which are of mixed type, perhaps type 1 for certain degrees of freedom and type 2 for other degrees of freedom. However, the general features are now clear, and it is more important to consider whether every canonical transformation is obtainable from a generating function of *some* type. The answer is yes. This is assured by the fact that it is always possible to choose *some* set of f variables from the set of 2f new coordinates and momenta $(Q, P)$ so that, together with the f old coordinates q, we obtain 2f independent variables.[3]

[3] C. Caratheodory, *Calculus of Variations and Partial Differential Equations of the First Order, Part 1: Partial Differential Equations of the First Order*, (Holden-Day, San Francisco, 1965), trans. Robert B. Dean and Julius J. Brandstatter, p. 92; V.I. Arnol'd, *Mathematical Methods of Classical Mechanics*, (Springer-Verlag New York, Inc. 1978), trans. K. Vogtmann and A. Weinstein, section 48.

## Identity and point transformations

We have already noted that the identity transformation is a canonical transformation, and that a type 1 generating function does not work. What then is a suitable generating function? It is easy to see that the type 2 generating function

$$F_{identity} = \sum_{a=1}^{f} q_a P_a$$

does the job. Suppose now that we replace the old coordinates $q_a$ in $F_{identity}$ by a set of independent functions $f_a(q,t)$ of the old coordinates, obtaining

$$F(q,P,t) = \sum_{a=1}^{f} f_a(q,t) P_a .$$

What transformation does this generate? We readily calculate

$$Q_a = \frac{\partial F}{\partial P_a} = f_a(q,t) \qquad p_a = \frac{\partial F}{\partial q_a} = \sum_{b=1}^{f} \frac{\partial f_a}{\partial q_b} P_b .$$

This is a **point transformation**, in which the new coordinates are determined solely by the old coordinates (with time as a parameter), and is the type of transformation encountered in Lagrangian mechanics. It is now supplemented by an associated (linear) transformation of the momentum.

As specific examples, let us consider again the space-time transformations, as applied to a single particle of mass m.

(a) A **spatial displacement**,

$$\mathbf{r}' = \frac{\partial F}{\partial \mathbf{p}'} = \mathbf{r} + \mathbf{a},$$

is generated by $F(\mathbf{r},\mathbf{p}') = (\mathbf{r} + \mathbf{a}) \cdot \mathbf{p}'$. The associated transformation of the momentum is

$$\mathbf{p} = \frac{\partial F}{\partial \mathbf{r}} = \mathbf{p}';$$

that is, the momentum is the same in the two frames.

(b) A **spatial rotation** about the z-axis through an angle θ,

$$x' = \frac{\partial F}{\partial p'_x} = x\cos\theta - y\sin\theta, \quad y' = \frac{\partial F}{\partial p'_y} = x\sin\theta + y\cos\theta, \quad z' = \frac{\partial F}{\partial p'_z} = z,$$

is generated by

$$F = (x\cos\theta - y\sin\theta)p'_x + (x\sin\theta + y\cos\theta)p'_y + zp'_z.$$

The associated transformation of the momentum is

$$p_x = \frac{\partial F}{\partial x} = p'_x\cos\theta + p'_y\sin\theta, \quad p_y = \frac{\partial F}{\partial y} = -p'_x\sin\theta + p'_y\cos\theta, \quad p_z = \frac{\partial F}{\partial z} = p'_z.$$

These can be inverted to give

$$p'_x = p_x\cos\theta - p_y\sin\theta, \quad p'_y = p_x\sin\theta + p_y\cos\theta, \quad p'_z = p_z.$$

The components of the momentum transform in the same way as the coordinates; that is, momentum is a vector.

(c) A **Galilean transformation**

$$\mathbf{r}' = \frac{\partial F}{\partial \mathbf{p}'} = \mathbf{r} + \mathbf{v}t$$

can be generated by

$$F(\mathbf{r},\mathbf{p}',t) = (\mathbf{r} + \mathbf{v}t)\cdot\mathbf{p}'.$$

The associated transformation of the momentum is

$$\mathbf{p} = \frac{\partial F}{\partial \mathbf{r}} = \mathbf{p}',$$

so with the above generating function the momentum in the two frames is the same. This is not what we would expect for a Galilean transformation. Rather, since particles have an additional velocity $\mathbf{v}$ in the prime frame, we would expect a particle of mass m to have an additional momentum $m\mathbf{v}$, so that

$$\mathbf{p}' = \mathbf{p} + m\mathbf{v}.$$

We can achieve this, without affecting the coordinate transformation, by modifying the generating function, adding to the original a suitable function of $\mathbf{r}$ and t. In particular, if we take

$$F(\mathbf{r},\mathbf{p}',t) = (\mathbf{r} + \mathbf{v}t)\cdot\mathbf{p}' - m\mathbf{v}\cdot\mathbf{r} + f(t)$$

where $f(t)$ is any function of time, then both the transformation of the coordinate $\mathbf{r}$ and the transformation of the momentum $\mathbf{p}$ are as expected. There is now, however, a feature which distinguishes this transformation from the other space-time transformations: the generating function depends not only on the frames, but also (because of the parameter m) on the system being considered.

## Infinitesimal canonical transformations

We have seen that the identity transformation is a canonical transformation, generated by the type 2 generating function $F_{\text{identity}} = \sum_{a=1}^{f} q_a P_a$. We now consider transformations "near" the identity. These are generated by functions "near" $F_{\text{identity}}$. We set

$$F(q,P,t) = \sum_{a=1}^{f} q_a P_a + \varepsilon G(q,P,t)$$

where G is any function, and the infinitesimal ε serves to remind us that the change is "small," so that terms of order $\varepsilon^2$ and higher can be dropped. F generates the transformation

$$Q_a = \frac{\partial F}{\partial P_a} = q_a + \varepsilon \frac{\partial G(q,P,t)}{\partial P_a} \qquad p_a = \frac{\partial F}{\partial q_a} = P_a + \varepsilon \frac{\partial G(q,P,t)}{\partial q_a}.$$

Since the term involving G is already first order in ε, we can replace the new momentum P in G by the old momentum p, the difference affecting only terms of order $\varepsilon^2$ and higher. This gives the **infinitesimal canonical transformation** in explicit form

$$Q_a = q_a + \varepsilon \frac{\partial G(q,p,t)}{\partial p_a} \qquad P_a = p_a - \varepsilon \frac{\partial G(q,p,t)}{\partial q_a}.$$

We refer to εG as the **generator** of the infinitesimal canonical transformation. Thus, for example, the generator for an infinitesimal spatial displacement $\boldsymbol{\delta a}$ is $\mathbf{p}\cdot\boldsymbol{\delta a}$ where $\mathbf{p}$ is the linear momentum; that for an infinitesimal rotation $\delta\theta$ about the z-axis is $(xp_y - yp_x)\delta\theta = L_z\,\delta\theta$ where $L_z$ is the z-component of the angular momentum; and that for a Galilean transformation $\boldsymbol{\delta v}$ is $(\mathbf{p}t - m\mathbf{r})\cdot\boldsymbol{\delta v}$. These can be obtained from the generating functions for the finite transformations by replacing the parameters with infinitesimals and retaining only zeroth and first order terms.

An especially important case is the following. Let G be the Hamiltonian H, and let ε be dt, an infinitesimal increase in time. For this choice the new variables are

$$Q_a(t) = q_a(t) + \frac{\partial H}{\partial p_a}dt = q_a(t) + \frac{dq_a}{dt}dt = q_a(t+dt)$$

$$P_a(t) = p_a(t) - \frac{\partial H}{\partial q_a}dt = p_a(t) + \frac{dp_a}{dt}dt = p_a(t+dt)\ .$$

From the active point of view, this represents a displacement of points in phase space. The displacement is that which occurs as the system develops in time from $t$ to $t + dt$. Put another way, we can view the infinitesimal time development as the result of an infinitesimal canonical transformation. Since the finite time development from $t_0$ to $t$ can be built up from a succession of infinitesimal developments, each of which can be described by an infinitesimal canonical transformation, and since a succession of canonical transformations is again a canonical transformation (canonical transformations form a group), the finite time development from $t_0$ to $t$ can be viewed as the result of a finite canonical transformation. The variables at time $t$ are related to the initial variables at time $t_0$ by a canonical transformation. Finding this canonical transformation (that is, its generating function) amounts to solving the dynamical problem. We return to this idea in Chapter VIII.

## Invariance transformations

In a canonical transformation $(q,p) \to (Q,P)$ the canonical equations retain their general form (that is, are covariant), the old set

$$\frac{dq_a}{dt} = \frac{\partial H}{\partial p_a} \qquad \frac{dp_a}{dt} = -\frac{\partial H}{\partial q_a}$$

going over to the new set

$$\frac{dQ_a}{dt} = \frac{\partial K}{\partial P_a} \qquad \frac{dP_a}{dt} = -\frac{\partial K}{\partial Q_a}$$

with new Hamiltonian given by

$$K(Q,P,t) = H(q,p,t) + \frac{\partial F}{\partial t}$$

where $F$ is the generating function of the transformation. Since $K$ is usually a different function of the new variables $(Q,P)$ than $H$ is of the old variables $(q,p)$, the explicit form of the right-hand side of the canonical equations is usually different for the new and old variables. However, if $K$ happens to be the same function of the new variables as $H$ was of the old variables, that is if

$$K(Q,P,t) = H(Q,P,t),$$

the explicit form of the canonical equations *is* the same for the old and new variables. We then say that the system is **invariant** under the transformation, and that the transformation is an **invariance transformation**. Putting the above two relations together, we have the condition for invariance

$$H(Q,P,t) = H(q,p,t) + \frac{\partial F}{\partial t}.$$

An important case is the **infinitesimal invariance transformation**. If we set

$$Q_a = q_a + \varepsilon \frac{\partial G(q,p,t)}{\partial p_a} \qquad P_a = p_a - \varepsilon \frac{\partial G(q,p,t)}{\partial q_a}$$

where $\varepsilon G$ is the generator of the infinitesimal canonical transformation, we have the condition

$$H(q + \varepsilon \frac{\partial G}{\partial p}, p - \varepsilon \frac{\partial G}{\partial q}, t) = H(q,p,t) + \varepsilon \frac{\partial G}{\partial t}.$$

If we expand and regroup this equation, we can write the condition for invariance in the form

$$\frac{dG}{dt} = \frac{\partial G}{\partial t} + [G,H] = 0.$$

The generator of the infinitesimal invariance transformation is thus a constant of the motion, and we can expand on the Chapter V result, now saying:

Associated with
any infinitesimal invariance transformation
is
a constant of the motion,
*which is the generator of the transformation.*

For example, the free particle system with Hamiltonian

$$H = \frac{|\mathbf{p}|^2}{2m}$$

is obviously invariant under spatial displacement and spatial rotation; the associated constants of the motion are proportional to the linear and angular momenta. Less obviously, it is also invariant under Galilean transformation. The new Hamiltonian is given by

$$K = H + \partial F/\partial t$$

where $F(\mathbf{r},\mathbf{p}',t) = (\mathbf{r} + \mathbf{v}t)\cdot\mathbf{p}' - m\mathbf{v}\cdot\mathbf{r} + f(t)$ is the generating function of the Galilean transformation. We thus find

$$K = \frac{|\mathbf{p}' - m\mathbf{v}|^2}{2m} + \mathbf{v}\cdot\mathbf{p}' + \frac{df(t)}{dt} = \frac{|\mathbf{p}'|^2}{2m} + \frac{1}{2}mv^2 + \frac{df(t)}{dt}.$$

So if we take $f(t) = -\frac{1}{2}mv^2t$, the new Hamiltonian is $K = \frac{|\mathbf{p}'|^2}{2m}$ and we have invariance. Actually, since the last two terms in K are independent of the canonical variables, being at most a function of time, their presence or absence does not affect the role of K as Hamiltonian; the dynamical equations are the same either way. It is, however, neater to eliminate them as we have done. The full generating function of a Galilean transformation is then

$$F(\mathbf{r},\mathbf{p}',t) = (\mathbf{r} + \mathbf{v}t)\cdot\mathbf{p}' - m\mathbf{v}\cdot\mathbf{r} - \tfrac{1}{2}mv^2t.$$

The generator for the infinitesimal Galilean transformation is $(\mathbf{p}t - m\mathbf{r})\cdot\delta\mathbf{v}$. It is a constant of the motion for the free particle system. The term in brackets can be identified as $-m\mathbf{r}_0$ where $\mathbf{r}_0$ is the initial position of the particle.

## Lagrange and Poisson brackets

In our discussion to this point, canonical transformations have been intimately connected with dynamics, with the time development of a system. Indeed, they were introduced as those transformations (with certain non-scale-changing requirements) which preserve the form of the dynamical equations, that is under which Hamilton's canonical equations are covariant. This view is somewhat misleading. While the connection of canonical transformations to dynamics is important, we should take a broader perspective, viewing canonical transformations simply as a class of transformation on phase space without reference to dynamics. After all, we have already noted that any *particular* canonical transformation leaves the form of the canonical equations unaffected, independent of the form of the Hamiltonian; a canonical transformation is not tied to any specific dynamical system. Canonical transformations may still, of course, be "time dependent," but the "time" in the transformation is to be regarded simply as a continuous parameter on which the form of the transformation depends. We note that the definition of a canonical transformation really contains two ideas. In the first place, there is the equation arising from the variation of the canonical variables, "covariance of the **canonical one-form** $\sum_{c=1}^{f} p_c\delta q_c$ up to a total differential," thus

$$\sum_{c=1}^{f}(p_c\delta q_c - P_c\delta Q_c) = \delta F.$$

Here the time t, if it appears in the transformation equations, is assumed to be held fixed; hence the "δ" rather than "d." In the second place, there is the equation arising from the variation of the time, the canonical variables being held fixed. This gives the new Hamiltonian K. For now we consider only the above equation, returning in the next section to the dynamical aspects and, in particular, to the existence of a new Hamiltonian.

If we take F as a function of one of the mixed sets of 2f variables, (q,Q) or (q,P) for example, we recover the generating function description of canonical transformations.

Let us now see what happens if we take F as a function of a more natural set of 2f independent variables, the new canonical variables $(Q, P)$. We write the left-hand side of the preceding equation as

$$\sum_{a=1}^{f}\left[\left(\sum_{c=1}^{f} p_c \frac{\partial q_c}{\partial Q_a} - P_a\right)\delta Q_a + \left(\sum_{c=1}^{f} p_c \frac{\partial q_c}{\partial P_a}\right)\delta P_a\right].$$

The necessary and sufficient conditions for this to be an exact differential are that the cross-derivatives be equal,

$$\frac{\partial}{\partial Q_b}\left(\sum_{c=1}^{f} p_c \frac{\partial q_c}{\partial Q_a} - P_a\right) = \frac{\partial}{\partial Q_a}\left(\sum_{c=1}^{f} p_c \frac{\partial q_c}{\partial Q_b} - P_b\right)$$

$$\frac{\partial}{\partial P_b}\left(\sum_{c=1}^{f} p_c \frac{\partial q_c}{\partial Q_a} - P_a\right) = \frac{\partial}{\partial Q_a}\left(\sum_{c=1}^{f} p_c \frac{\partial q_c}{\partial P_b}\right)$$

$$\frac{\partial}{\partial P_b}\left(\sum_{c=1}^{f} p_c \frac{\partial q_c}{\partial P_a}\right) = \frac{\partial}{\partial P_a}\left(\sum_{c=1}^{f} p_c \frac{\partial q_c}{\partial P_b}\right).$$

On expanding these, we find

$$\sum_{c=1}^{f}\left(\frac{\partial q_c}{\partial Q_a}\frac{\partial p_c}{\partial Q_b} - \frac{\partial p_c}{\partial Q_a}\frac{\partial q_c}{\partial Q_b}\right) = 0$$

$$\sum_{c=1}^{f}\left(\frac{\partial q_c}{\partial Q_a}\frac{\partial p_c}{\partial P_b} - \frac{\partial p_c}{\partial Q_a}\frac{\partial q_c}{\partial P_b}\right) = \delta_{ab}$$

$$\sum_{c=1}^{f}\left(\frac{\partial q_c}{\partial P_a}\frac{\partial p_c}{\partial P_b} - \frac{\partial p_c}{\partial P_a}\frac{\partial q_c}{\partial P_b}\right) = 0\,.$$

The sums on the left-hand sides all have similar structure and can be expressed in terms of a new kind of quantity called a Lagrange bracket. We suppose that the canonical variables $(q, p)$ are functions of two variables, say u and v. The **Lagrange bracket** of u and v with respect to $(q, p)$ is then defined as

$$\{u, v\}_{q,p} = \sum_{c=1}^{f}\left(\frac{\partial q_c}{\partial u}\frac{\partial p_c}{\partial v} - \frac{\partial p_c}{\partial u}\frac{\partial q_c}{\partial v}\right).$$

The necessary and sufficient conditions for a transformation $(q, p) \rightarrow (Q, P)$ to be canonical can now be expressed in terms of Lagrange brackets, thus

$$\{Q_a, Q_b\}_{q,p} = 0 \qquad \{Q_a, P_b\}_{q,p} = \delta_{ab} \qquad \{P_a, P_b\}_{q,p} = 0.$$

The u and v are here the new canonical variables $(Q,P)$ taken a pair at a time.

Lagrange brackets look like reciprocals of Poisson brackets; the derivatives are, as it were, "upside-down." Recall that for a Poisson bracket we suppose that we have two functions, say u and v, of the canonical variables $(q,p)$. The **Poisson bracket** of u and v with respect to $(q,p)$ is defined as

$$[u,v]_{q,p} = \sum_{d=1}^{f}\left(\frac{\partial u}{\partial q_d}\frac{\partial v}{\partial p_d} - \frac{\partial u}{\partial p_d}\frac{\partial v}{\partial q_d}\right).$$

To see the connection between Lagrange brackets and Poisson brackets, let $u_k(q,p,t)$, where $k = 1,2,\cdots,2f$, be any set of 2f independent functions of the 2f canonical variables $(q,p)$, and possibly also of the time t as a parameter. These can be inverted to express the canonical variables as functions of the u's. If we imagine the canonical variables as functions of any two of the u's, say $u_k$ and $u_i$, we can form a Lagrange bracket $\{u_k,u_i\}_{q,p}$, and if we imagine any two of the u's, say $u_k$ and $u_j$, as functions of the canonical variables, we can form a Poisson bracket $[u_k,u_j]_{q,p}$. Now consider

$$\sum_{k=1}^{2f}\{u_k,u_i\}[u_k,u_j] = \sum_{k=1}^{2f}\sum_{c=1}^{f}\sum_{d=1}^{f}\left(\frac{\partial q_c}{\partial u_k}\frac{\partial p_c}{\partial u_i} - \frac{\partial p_c}{\partial u_k}\frac{\partial q_c}{\partial u_i}\right)\left(\frac{\partial u_k}{\partial q_d}\frac{\partial u_j}{\partial p_d} - \frac{\partial u_k}{\partial p_d}\frac{\partial u_j}{\partial q_d}\right).$$

On expanding this, we encounter

$$\sum_{k=1}^{2f}\frac{\partial q_c}{\partial u_k}\frac{\partial u_k}{\partial q_d} = \delta_{cd},\quad \sum_{k=1}^{2f}\frac{\partial q_c}{\partial u_k}\frac{\partial u_k}{\partial p_d} = 0,\quad \sum_{k=1}^{2f}\frac{\partial p_c}{\partial u_k}\frac{\partial u_k}{\partial q_d} = 0,\ \text{and}\ \sum_{k=1}^{2f}\frac{\partial p_c}{\partial u_k}\frac{\partial u_k}{\partial p_d} = \delta_{cd}.$$

We thus get

$$\sum_{k=1}^{2f}\{u_k,u_i\}[u_k,u_j] = \sum_{c=1}^{f}\sum_{d=1}^{f}\left(\frac{\partial p_c}{\partial u_i}\frac{\partial u_j}{\partial p_d} + \frac{\partial q_c}{\partial u_i}\frac{\partial u_j}{\partial q_d}\right)\delta_{cd} = \sum_{c=1}^{f}\left(\frac{\partial p_c}{\partial u_i}\frac{\partial u_j}{\partial p_c} + \frac{\partial q_c}{\partial u_i}\frac{\partial u_j}{\partial q_c}\right) = \delta_{ij},$$

which expresses quantitatively the reciprocal nature of the brackets. In particular, if we arrange the sets of Lagrange and Poisson brackets in matrix form,

$$\begin{vmatrix} \{u_1,u_1\} & \{u_2,u_1\} & \cdots \\ \{u_1,u_2\} & \{u_2,u_2\} & \\ \vdots & & \end{vmatrix}\begin{vmatrix} [u_1,u_1] & [u_1,u_2] & \cdots \\ [u_2,u_1] & [u_2,u_2] & \\ \vdots & & \end{vmatrix} = \begin{vmatrix} 1 & 0 & \cdots \\ 0 & 1 & \\ \vdots & & \end{vmatrix},$$

we see that the transpose of the matrix of the Lagrange brackets and the matrix of the Poisson brackets are inverses of one another. Now suppose that the u's are related to the $(q,p)$ by a canonical transformation, that they form a set of new canonical variables with

$$u_a = Q_a \qquad u_{a+f} = P_a \qquad \text{where } a = 1,2,\cdots,f.$$

We then know the Lagrange brackets and can use the above relation to find the Poisson brackets

$$[Q_a,Q_b]_{q,p} = 0 \qquad [Q_a,P_b]_{q,p} = \delta_{ab} \qquad [P_a,P_b]_{q,p} = 0 .$$

These are the necessary and sufficient conditions for a transformation to be canonical, now expressed in terms of Poisson brackets. They form a convenient way to test whether or not a given transformation $(q,p) \rightarrow (Q,P)$ is canonical.

We notice that since the Poisson brackets of the new variables $(Q,P)$ with respect to the new variables are obviously given by

$$[Q_a,Q_b]_{Q,P} = 0 \qquad [Q_a,P_b]_{Q,P} = \delta_{ab} \qquad [P_a,P_b]_{Q,P} = 0,$$

the fundamental Poisson brackets of the new variables are the same whether we use the old variables or the new variables to evaluate them; they are invariant under canonical transformations. We now show that this applies to all Poisson brackets. Consider two arbitrary functions $u(q,p,t)$ and $v(q,p,t)$ of the canonical variables $(q,p)$ and possibly also of the time t, and form their Poisson bracket $[u,v]_{q,p}$. We wish to see how the Poisson bracket transforms if we transform to new canonical variables $(Q,P)$. To help keep the variables straight, introduce $\bar{u}(Q,P,t) = u(q(Q,P,t),p(Q,P,t),t)$ and $\bar{v}(Q,P,t) = v(q(Q,P,t),p(Q,P,t),t)$, so $\bar{u}$ and u have the same value at a given point in phase space, viewing the transformation passively, but their functional forms are different. Using the chain rule for partial derivatives, we have

$$[u,v]_{q,p} = \sum_{a=1}^{f}\left(\frac{\partial u}{\partial q_a}\frac{\partial v}{\partial p_a} - \frac{\partial u}{\partial p_a}\frac{\partial v}{\partial q_a}\right)$$
$$= \sum_{a=1}^{f}\sum_{b=1}^{f}\sum_{c=1}^{f}\left[\left(\frac{\partial \bar{u}}{\partial Q_b}\frac{\partial Q_b}{\partial q_a} + \frac{\partial \bar{u}}{\partial P_b}\frac{\partial P_b}{\partial q_a}\right)\left(\frac{\partial \bar{v}}{\partial Q_c}\frac{\partial Q_c}{\partial p_a} + \frac{\partial \bar{v}}{\partial P_c}\frac{\partial P_c}{\partial p_a}\right) - \begin{pmatrix} q_a \leftrightarrow p_a \\ \text{interchange}\end{pmatrix}\right] .$$

Expanding this and collecting similar terms, we find that the Poisson bracket can be written

$$[u,v]_{q,p} = \sum_{b=1}^{f}\sum_{c=1}^{f}\left(\frac{\partial \bar{u}}{\partial Q_b}\frac{\partial \bar{v}}{\partial Q_c}[Q_b,Q_c]_{q,p} + \frac{\partial \bar{u}}{\partial Q_b}\frac{\partial \bar{v}}{\partial P_c}[Q_b,P_c]_{q,p}\right.$$
$$\left. + \frac{\partial \bar{u}}{\partial P_b}\frac{\partial \bar{v}}{\partial Q_c}[P_b,Q_c]_{q,p} + \frac{\partial \bar{u}}{\partial P_b}\frac{\partial \bar{v}}{\partial P_c}[P_b,P_c]_{q,p}\right) .$$

Now use the Poisson bracket conditions, and find

$$[u,v]_{q,p} = \sum_{b=1}^{f}\left(\frac{\partial \bar{u}}{\partial Q_b}\frac{\partial \bar{v}}{\partial P_b} - \frac{\partial \bar{u}}{\partial P_b}\frac{\partial \bar{v}}{\partial Q_b}\right) = [\bar{u},\bar{v}]_{Q,P} .$$

We have thus shown that an arbitrary Poisson bracket is invariant under canonical transformations; it can be evaluated using any set of canonical variables. This is yet another way to state the conditions that a transformation be canonical.

## Time dependence

Now turn to the time dependence of the variables. If we imagine the new canonical variables as functions of the old, then

$$\frac{dQ_a}{dt} = [Q_a, H]_{q,p} + \left(\frac{\partial Q_a}{\partial t}\right)_{q,p} \qquad \frac{dP_a}{dt} = [P_a, H]_{q,p} + \left(\frac{\partial P_a}{\partial t}\right)_{q,p} .$$

Since the Poisson bracket is invariant, we can set

$$[Q_a, H]_{q,p} = [Q_a, \overline{H}]_{Q,P} = \frac{\partial \overline{H}}{\partial P_a} \qquad [P_a, H]_{q,p} = [P_a, \overline{H}]_{Q,P} = -\frac{\partial \overline{H}}{\partial Q_a} .$$

Thus, if the canonical transformation does not depend explicitly on the time, the new variables satisfy the canonical equations of motion with Hamiltonian $\overline{H}$. What if the transformation *does* depend explicitly on the time? We must then consider what to do with the partial time derivatives of Q and P. To find these, we return to the definition of a canonical transformation,

$$\sum_{a=1}^{f} (p_a \delta q_a - P_a \delta Q_a) = \delta F ,$$

and differentiate with respect to t, holding $(q,p)$ fixed. The result is

$$\sum_{a=1}^{f} \left[ -\left(\frac{\partial P_a}{\partial t}\right)_{q,p} \delta Q_a - P_a \delta \left(\frac{\partial Q_a}{\partial t}\right)_{q,p} \right] = \delta \left(\frac{\partial F}{\partial t}\right)_{q,p} .$$

This can be written

$$\sum_{a=1}^{f} \left[ -\left(\frac{\partial P_a}{\partial t}\right)_{q,p} \delta Q_a + \left(\frac{\partial Q_a}{\partial t}\right)_{q,p} \delta P_a \right] = \delta G$$

where

$$G = \left(\frac{\partial F}{\partial t}\right)_{q,p} + \sum_{a=1}^{f} P_a \left(\frac{\partial Q_a}{\partial t}\right)_{q,p} .$$

If we take G as a function of the set of 2f independent variables $(Q,P)$, we have

$$\delta G = \sum_{a=1}^{f}\left(\frac{\partial G}{\partial Q_a}\delta Q_a + \frac{\partial G}{\partial P_a}\delta P_a\right).$$

Equating the coefficients of the $\delta Q$'s and the $\delta P$'s on the two sides of the above equation then gives the required partial time derivatives of the Q's and P's,

$$\left(\frac{\partial Q_a}{\partial t}\right)_{q,p} = \frac{\partial G}{\partial P_a} \qquad \left(\frac{\partial P_a}{\partial t}\right)_{q,p} = -\frac{\partial G}{\partial Q_a}.$$

If these expressions are now inserted into the equations at the beginning of this section, we see that in general the new variables $(Q,P)$ satisfy canonical equations of motion with new Hamiltonian $\overline{H} + G$.

Note that the "t" in the above equations does not have to be the "time," but could be *any parameter* on which the transformation $Q = Q(q,p,t)$, $P = P(q,p,t)$ depends. These are, in fact, the equations of an arbitrary infinitesimal canonical transformation with generator $G\,dt$.

The function G is simply related to the generating function of the canonical transformation. For a type 1 generating function, $F = F_1(q,Q,t)$, and

$$\left(\frac{\partial F}{\partial t}\right)_{q,p} = \left(\frac{\partial F_1}{\partial t}\right)_{q,Q} + \sum_{a=1}^{f}\left(\frac{\partial F_1}{\partial Q_a}\right)_{q,t}\left(\frac{\partial Q_a}{\partial t}\right)_{q,p} = \left(\frac{\partial F_1}{\partial t}\right)_{q,Q} - \sum_{a=1}^{f} P_a\left(\frac{\partial Q_a}{\partial t}\right)_{q,p},$$

where we have made use of the canonical transformation equation $P = -\partial F_1/\partial Q$. So we have

$$G = \left(\frac{\partial F_1}{\partial t}\right)_{q,Q}.$$

For a type 2 generating function, $F = F_2(q,P,t) - \sum_{a=1}^{f} Q_a P_a$, and

$$\left(\frac{\partial F}{\partial t}\right)_{q,p} = \left(\frac{\partial F_2}{\partial t}\right)_{q,P} + \sum_{a=1}^{f}\left[\left(\frac{\partial F_2}{\partial P_a}\right)_{q,t}\left(\frac{\partial P_a}{\partial t}\right)_{q,p} - Q_a\left(\frac{\partial P_a}{\partial t}\right)_{q,p} - P_a\left(\frac{\partial Q_a}{\partial t}\right)_{q,p}\right]$$
$$= \left(\frac{\partial F_2}{\partial t}\right)_{q,P} - \sum_{a=1}^{f} P_a\left(\frac{\partial Q_a}{\partial t}\right)_{q,p},$$

where we have made use of the canonical transformation equation $Q = \partial F_2/\partial P$. So we have

$$G = \left(\frac{\partial F_2}{\partial t}\right)_{q,P}.$$

In either case G is the partial time derivative of the generating function with the appropriate set of mixed variables held fixed. This then returns us to our earlier expressions for the Hamiltonian K for the new variables, namely

$$K = H + \frac{\partial F_{1,2}}{\partial t}.$$

## Integral invariants

We have seen that a transformation from variables (q,p) to variables (Q,P) is canonical if it preserves the **canonical one-form** $p\delta q$ up to a total differential, thus[4]

$$p\delta q - P\delta Q = \delta F.$$

We assume that the old and new momenta are single-valued "functions" of the state of the system (but the old and new coordinates need not be). The function F is then also single-valued, so if we integrate along any curve in phase space which starts at a given state and returns to the same state, the right-hand side gives zero and we have

$$\oint p\delta q = \oint P\delta Q.$$

We see that this integral, called the **circulation**, is invariant under canonical transformations. It should be emphasized that the curves we are here considering have nothing to do with dynamical trajectories; they are arbitrary "closed" curves in phase space. Indeed, dynamics (time) has yet to enter the picture.

We can express the invariance of circulation differently by making use of **Stokes' theorem**. This says

$$\oint p\delta q = \iint(\delta_1 q\,\delta_2 p - \delta_2 q\,\delta_1 p).$$

The integral on the left is around a simple closed curve in phase space, whereas that on the right is over a simply-connected two-dimensional surface which has the curve as its edge. Basically, Stokes' theorem states the equivalence of two ways of evaluating the area in the q-p plane enclosed by the curve (Fig. 7.01). On the left-hand side we add up the areas $p\delta q$ of little rectangles of height p and width $\delta q$, whereas on the right-hand side we add up the areas $\delta_1 q\,\delta_2 p - \delta_2 q\,\delta_1 p$ of little parallelograms with adjacent sides $(\delta_1 q, \delta_1 p)$ and $(\delta_2 q, \delta_2 p)$.

[4]In this section we use an abbreviated notation, suppressing the summation index and the summation sign. Thus $p\delta q$ stands for $\sum_{a=1}^{f} p_a\,\delta q_a$.

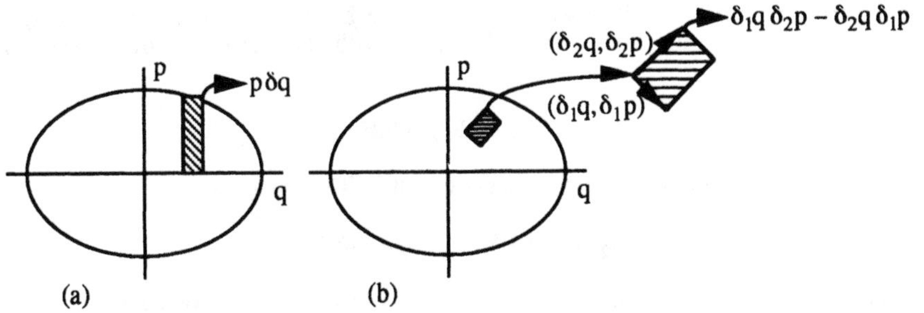


Fig. 7.01. Area in phase plane: (a) as rectangles; (b) as parallelograms

In particular, the circulation around one of these infinitesimal parallelograms is (Fig. 7.02)

$$\begin{aligned}\oint p\,\delta q &= (p\,\delta q)_a + (p\,\delta q)_b + (p\,\delta q)_c + (p\,\delta q)_d \\ &= \delta_2(p\,\delta_1 q) - \delta_1(p\,\delta_2 q) \\ &= \delta_1 q\,\delta_2 p - \delta_2 q\,\delta_1 p\ .\end{aligned}$$

The second equality follows from grouping the first term with the third term and the second term with the fourth term in the preceding line; the third equality then follows from $\delta_2\delta_1 q = \delta_1\delta_2 q$.

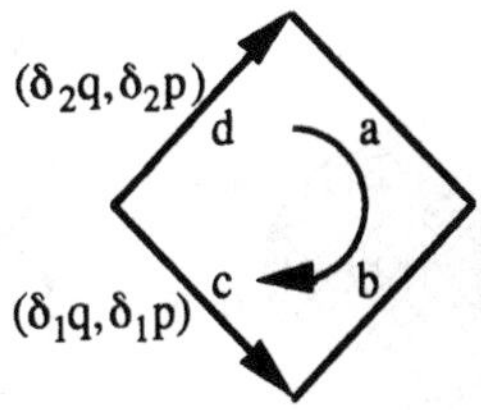


Fig. 7.02 Circulation around infinitesimal parallelogram

The last expression $\delta_1 q\delta_2 p - \delta_2 q\delta_1 p$ is called the **canonical two-form**. It is determined by two infinitesimal displacements in phase space and may be thought of as a little patch of area.[5] Since the circulation is invariant under canonical transformations, the canonical two-form is also invariant,

$$\delta_1 q\,\delta_2 p - \delta_2 q\,\delta_1 p = \delta_1 Q\delta_2 P - \delta_2 Q\delta_1 P;$$

it does not matter which set of canonical variables we use in evaluating the canonical two-form. This is yet another way to characterize a canonical transformation. The result is

[5]More fully, the canonical two-form is $\sum_{a=1}^{f}(\delta_1 q_a\,\delta_2 p_a - \delta_2 q_a\,\delta_1 p_a)$ and is the sum of the areas (with appropriate + or - sign) of the projections of the parallelogram onto the various degrees of freedom.

sufficiently important that it is worthwhile to give an alternate and perhaps simpler derivation which avoids integration. We start with the definition of a canonical transformation,

$$p\delta_1 q - P\delta_1 Q = \delta_1 F$$

with displacement 1, and subject it to a displacement 2. The change is

$$\delta_2(p\delta_1 q) - \delta_2(P\delta_1 Q) = \delta_2\delta_1 F.$$

We now interchange the displacements 1 and 2 and subtract the two results. Since $\delta_2\delta_1 F = \delta_1\delta_2 F$, the right-hand side gives zero and the left-hand side can be rearranged to give

$$\delta_2(p\delta_1 q) - \delta_1(p\delta_2 q) = \delta_2(P\delta_1 Q) - \delta_1(P\delta_2 Q).$$

On expanding this result, we recover invariance of the canonical two-form.

A convenient way to describe the surface over which the above integration is performed is to label points on it by two parameters, u and v, in terms of which the canonical variables can be written $q = q(u,v)$ and $p = p(u,v)$. A typical infinitesimal parallelogram is then the area enclosed by the curves u constant, $u + \delta u$ constant, v constant, and $v + \delta v$ constant (Fig. 7.03), and we can take $\delta_1 = \delta u(\partial/\partial u)$ and $\delta_2 = \delta v(\partial/\partial v)$.

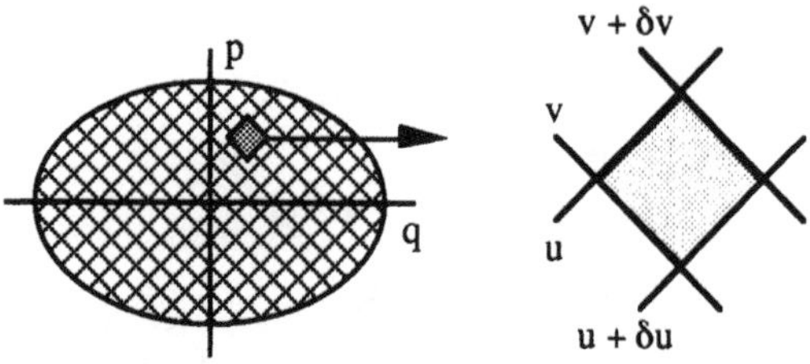


Fig. 7.03. Parametrizing the surface

The integral over the surface becomes

$$\iint \left( \frac{\partial q}{\partial u}\frac{\partial p}{\partial v} - \frac{\partial q}{\partial v}\frac{\partial p}{\partial u} \right) \delta u\, \delta v = \iint \{u,v\}_{q,p}\, \delta u\, \delta v$$

where $\{u,v\}_{q,p}$ is the Lagrange bracket of u and v with respect to the canonical variables q and p. We see that invariance of the canonical two-form under canonical transformations is another way of stating the invariance of the Lagrange brackets,

$$\{u,v\}_{q,p} = \{u,v\}_{Q,P}.$$

Let us now turn to dynamics, to the way things change with time. It is convenient to think of this as inducing an active transformation on phase space. Then, in the time interval t to $t + dt$, a point $(q, p)$ moves to $(q + dq, p + dp)$, where

$$dq = \frac{\partial H(q,p,t)}{\partial p} dt \qquad dp = -\frac{\partial H(q,p,t)}{\partial q} dt .$$

A neighboring point $(q + \delta q, p + \delta p)$ moves similarly

$$d(q + \delta q) = \frac{\partial H(q + \delta q, p + \delta p, t)}{\partial p} dt \qquad d(p + \delta p) = -\frac{\partial H(q + \delta q, p + \delta p, t)}{\partial q} dt .$$

So, expanding the right-hand side of the above with the aid of Taylor's theorem and subtracting $(dq, dp)$, we obtain the change in the little displacement $(\delta q, \delta p)$ in the time interval t to $t + dt$,

$$d\,\delta q = \left( \frac{\partial^2 H}{\partial q \partial p} \delta q + \frac{\partial^2 H}{\partial p \partial p} \delta p \right) dt \qquad d\,\delta p = -\left( \frac{\partial^2 H}{\partial q \partial q} \delta q + \frac{\partial^2 H}{\partial p \partial q} \delta p \right) dt .$$

From these basic relations we can find the changes in other quantities. In particular, the change in the canonical two-form in the time interval t to $t + dt$ is given by

$$d(\delta_1 q\, \delta_2 p - \delta_2 q\, \delta_1 p) = (d\,\delta_1 q) \delta_2 p + \delta_1 q (d\,\delta_2 p) - (1 \leftrightarrow 2).$$

Substituting for $d\,\delta q$ and for $d\,\delta p$, we find that all terms on the right cancel independent of the form of the Hamiltonian, so the change in the canonical two-form is *zero*. Hamiltonian dynamics is such that the canonical two-form is constant in time,[6]

$$\delta_1 q\, \delta_2 p - \delta_2 q\, \delta_1 p = \text{constant} .$$

If we think of the canonical two-form as a little patch of area in phase space, then as time goes on the patch moves and distorts, but its area remains the same. More generally, if we consider a two-dimensional surface in phase space, as time goes on this surface moves and distorts, but its area $\iint (\delta_1 q\, \delta_2 p - \delta_2 q\, \delta_1 p)$ remains the same. By Stokes' theorem this surface integral is equal to the circulation $\oint p\, \delta q$ around the edge of the surface, so this shows also that the circulation around any simple co-moving curve[7] in phase space is constant in time.

---

[6]This also follows from the fact that the canonical two-form is invariant under canonical transformations, and that the canonical variables at time $t + dt$ are related to those at time t by a canonical transformation.
[7]A curve which moves with the phase points.

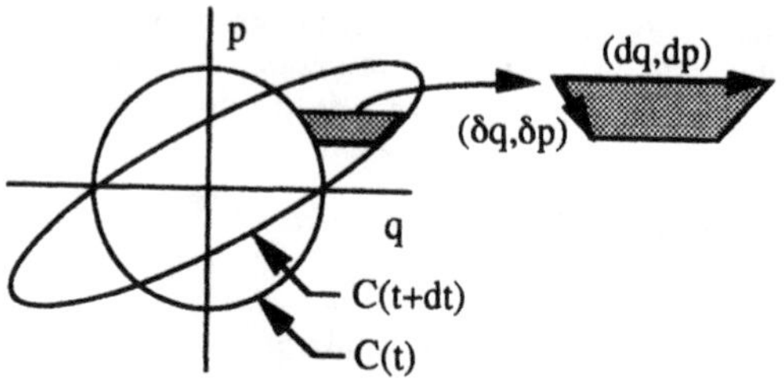


Fig. 7.04. Change in area

Another way to see this last result is as follows. From Fig. 7.04 the change in the area enclosed by the curve $C(t)$, in the time interval t to $t + dt$, is given by

$$\text{Change in area} = \oint(\delta q\, dp - dq\, \delta p),$$

where the integration (with respect to $(\delta q, \delta p)$) is around $C(t)$. Hamilton's canonical equations then imply that

$$\text{Change in area} = \oint[-\delta q\,(\partial H/\partial q)dt - (\partial H/\partial p)dt\,\delta p] = -\oint \delta H\, dt = 0,$$

since the change in the Hamiltonian in going around $C(t)$ is zero.

The canonical two-form

$$\omega(1,2) = \delta_1 q\, \delta_2 p - \delta_2 q\, \delta_1 p,$$

which is based on two independent displacements in phase space, is the first member of a family of invariants, collectively called **Poincaré integral invariants**. With its aid we can construct a four-form by taking the **exterior product** $\omega \wedge \omega$ of the two-form $\omega$ with itself,[8]

$$\omega \wedge \omega(1,2,3,4) = 2\big[\omega(1,2)\omega(3,4) - \omega(1,3)\omega(2,4) + \omega(1,4)\omega(2,3)\big].$$

The four-form is based on four independent displacements and, like the two-form, is antisymmetric under interchange of any two of the displacements. Since it is built up from the invariant two-form, the four-form is also invariant. This process, forming successive exterior powers of $\omega$, can be continued until we run out of independent displacements. Since phase space is 2f dimensional, the last member of the family is the 2f-form

[8] V.I. Arnol'd, *Mathematical Methods of Classical Mechanics*, (Springer-Verlag New York, Inc. 1978), trans. K. Vogtmann and A. Weinstein, p.170.

$$\underbrace{\omega \wedge \omega \wedge \cdots \wedge \omega}_{\text{f terms}}(1,2,\cdots,2f) = f!\begin{vmatrix} \delta_1 q_1 & \delta_1 p_1 & \delta_1 q_2 & \cdots \\ \delta_2 q_1 & \delta_2 p_1 & & \\ \delta_3 q_1 & & & \\ \vdots & & & \end{vmatrix}.$$

It is proportional to the determinant in which the 2f rows are the components of the 2f independent displacements. This determinant is the volume of a 2f dimensional parallelepiped with edges being the 2f displacements. As with all members of the family, it is invariant. We thus see that Hamiltonian dynamics implies that the volume of any simply-connected co-moving region of phase space is constant in time. This is known as **Liouville's theorem.**

In studying a system with a large number of degrees of freedom, such as a liter of gas molecules at STP, we are usually unable to specify exactly the initial state of the system. The best we can do is to give the probability for finding the system in various regions of phase space. Thus let $\rho(q_0,p_0,t_0)\delta V_0$ be the probability for finding the system in the little phase space volume $\delta V_0$ surrounding $(q_0,p_0)$ at time $t_0$.[9] In the time interval $t_0$ to t, each point in the little volume moves in accordance with the canonical equations, the point $(q_0,p_0)$ moving to $(q,p)$, and the little volume $\delta V_0$ which surrounds $(q_0,p_0)$ moving and distorting to become a little volume $\delta V$ surrounding $(q,p)$. Since no system points enter or leave the little moving volume, the probability for finding the system in this volume $\delta V$ at time t is the same as that for finding it in $\delta V_0$ at time $t_0$,

$$\rho(q,p,t)\delta V = \rho(q_0,p_0,t_0)\delta V_0.$$

On the other hand, according to Liouville's theorem the volumes $\delta V$ and $\delta V_0$ themselves are equal, and thus the density, if we follow the moving phase space point, remains constant,

$$\rho(q,p,t) = \rho(q_0,p_0,t_0).$$

Regions of phase space move like an incompressible fluid. By considering a small interval of time, we can rewrite this as

$$\frac{d\rho}{dt} = \frac{\partial\rho}{\partial t} + [\rho,H] = 0.$$

This result, also sometimes called Liouville's theorem, is of fundamental importance in statistical mechanics.[10]

---

[9]Alternatively, we can imagine the system replaced by an **ensemble** of identical independent systems with differing initial conditions consistent with our imperfect knowledge; $\rho(q_0,p_0,t_0)\delta V_0$ is then the number of such systems in the region $\delta V_0$ at time $t_0$.

[10]L. D. Landau and E. M. Lifshitz, *Statistical Physics*, 2nd ed., (Pergamon Press, Oxford, UK, 1969), trans. J. B. Sykes and M. J. Kearsley, p. 9; Kerson Huang, *Statistical Mechanics*, 2nd ed., (John Wiley and Sons, New York, NY, 1987), p. 64.

## Exercises

1. The motion of a particle of mass m undergoing constant acceleration a in one dimension is described by

$$x = x_0 + \frac{p_0}{m}t + \frac{1}{2}at^2$$
$$p = p_0 + mat \quad .$$

Show that the transformation from present "old" variables $(x,p)$ to initial "new" variables $(x_0,p_0)$ is a canonical transformation
(a) by Poisson bracket test
(b) by finding (for $t \neq 0$) the type 1 generating function $F_1(x,x_0,t)$.

2. (a) Show that

$$Q = -p$$
$$P = q + Ap^2,$$

(where A is any constant) is a canonical transformation,
(i) by evaluating $[Q,P]_{q,p}$
(ii) by expressing $p\,dq - P\,dQ$ as an exact differential $dF(q,Q)$. Hence find the type 1 generating function of the transformation. To do this, you must first use the transformation equations to express p, P in terms of q, Q.
(b) Use the relation $F_2 = F_1 + PQ$ to find the type 2 generating function $F_2(q,P)$, and check your result by showing that $F_2$ indeed generates the transformation.

3. The Hamiltonian for a particle moving vertically in a uniform gravitational field g is

$$H = \frac{p^2}{2m} + mgq\,.$$

(a) Find the new Hamiltonian for new canonical variables Q, P as given in exercise 7.02. Show that we can eliminate Q from the Hamiltonian (make Q cyclic) by choosing the constant A appropriately.
(b) With this choice of A write down and solve Hamilton's equations for the new canonical variables, and then use the transformation equations to find the original variables q, p as functions of time.

4. (a) Show that

$$Q = q\cos\theta - \frac{p}{m\omega}\sin\theta$$
$$P = m\omega q\sin\theta + p\cos\theta$$

is a canonical transformation,
(i) by evaluating $[Q,P]_{q,p}$
(ii) by expressing $p\,dq - P\,dQ$ as an exact differential $dF_1(q,Q,t)$. Hence find the type 1 generating function of the transformation. To do this, you must first use the transformation equations to express p, P in terms of q, Q.
(b) Use the relation $F_2 = F_1 + PQ$ to find the type 2 generating function $F_2(q,P)$, and check your result by showing that $F_2$ indeed generates the transformation.

5. Suppose that the $(q,p)$ of exercise 7.04 are canonical variables for a simple harmonic oscillator with Hamiltonian

$$H = \frac{p^2}{2m} + \frac{1}{2}m\omega^2 q^2.$$

(a) Find the Hamiltonian $K(Q,P,t)$ for the new canonical variables $(Q,P)$, assuming that the parameter $\theta$ is some function of time. Show that we can choose $\theta(t)$ so that $K = 0$.
(b) With this choice of $\theta(t)$ solve the new canonical equations to find $(Q,P)$ as functions of time, and then use the transformation equations to find the original variables $(q,p)$ as functions of time.

6. (a) Show that the Hamiltonian for a simple harmonic oscillator is invariant under the canonical transformation of exercise 7.04 (for $\theta$ constant).
(b) Find the associated constant of the motion.

7. (a) What are the conditions on the "small" constants a, b, c, d, e, and f in order that

$$q = Q + aQ^2 + 2bQP + cP^2$$

$$p = P + dQ^2 + 2eQP + fP^2$$

be a canonical transformation to first order in small quantities?
(b) The Hamiltonian for a slightly anharmonic oscillator is

$$H = \frac{p^2}{2m} + \frac{1}{2}m\omega^2 q^2 + \beta q^3$$

where $\beta$ is "small." Perform a canonical transformation of the type given in part (a) and adjust the constants so that the new Hamiltonian H does not contain an anharmonic term to first order in small quantities, thus

$$\overline{H} = \frac{P^2}{2m} + \frac{1}{2}m\omega^2 Q^2 + \text{second order terms}.$$

(c) Write down and solve Hamilton's equations for the new canonical variables, and then use the transformation equations to find the solution to the anharmonic oscillator problem valid to first order in small quantities.

8. (a) Show that

$$x' = x + \frac{1}{m}p_z\tau \qquad p'_x = p_x - mg\tau$$

$$y' = y \qquad p'_y = p_y$$

$$z' = z + \frac{1}{m}p_x\tau - \frac{1}{2}g\tau^2 \qquad p'_z = p_z$$

(where $\tau$ is any constant) is a canonical transformation by finding the type 2 generating function $F_2(x,y,z,p'_x,p'_y,p'_z)$.
(b) Show that the Hamiltonian

$$H = \frac{1}{2m}(p_x^2 + p_y^2 + p_z^2) + mgz$$

for a projectile near the surface of the earth is invariant under the canonical transformation given in part (a), and find the associated constant of the motion.

9. (a) Show that

$$Q_1 = \frac{1}{\sqrt{2}}\left(q_1 + \frac{p_2}{m\omega}\right) \qquad P_1 = \frac{1}{\sqrt{2}}(p_1 - m\omega q_2)$$

$$Q_2 = \frac{1}{\sqrt{2}}\left(q_1 - \frac{p_2}{m\omega}\right) \qquad P_2 = \frac{1}{\sqrt{2}}(p_1 + m\omega q_2),$$

(where $m\omega$ is a constant) is a canonical transformation by Poisson bracket test. This requires evaluating *six* simple Poisson brackets.

(b) Find a generating function $F(q_1,q_2,Q_1,P_2)$ for this transformation, type 1 in the first degree of freedom and type 2 in the second degree of freedom.

10. (a) Let x denote a column matrix of the canonical variables $q_1,q_2,p_1,p_2$ for a system with two degrees of freedom, and consider a linear transformation

$$x \to x' = Mx$$

where M is a $4 \times 4$ matrix with constant elements. Use the Poisson bracket conditions to find the conditions on the elements of M in order that this be a canonical transformation.

(b) Show that these are equivalent to requiring that M satisfy the condition

$$MJ\tilde{M} = J.$$

Here $\tilde{M}$ is the transpose of M, and J is the matrix

$$J = \begin{bmatrix} 0 & 1 \\ -1 & 0 \end{bmatrix}$$

where "0" stands for the $2 \times 2$ zero matrix and "1" stands for the $2 \times 2$ unit matrix. Matrices M which satisfy the above condition are called **symplectic** matrices.

11. The dynamics of a system of interacting particles is governed by a Hamiltonian

$$H = \sum_{i=1}^{N} \frac{|\mathbf{p}_i|^2}{2m_i} + \frac{1}{2}\sum_{i=1}^{N}\sum_{j=1}^{N} V_{ij}(\mathbf{r}_i - \mathbf{r}_j).$$

Suppose we view this system from a uniformly accelerating coordinate frame

$$\mathbf{r}_i' = \mathbf{r}_i - \tfrac{1}{2}\mathbf{a}t^2.$$

Show that we can choose the canonical transformation connecting the two frames (that is, its type 2 generating function $F_2(\mathbf{r},\mathbf{p}',t)$) so that the Hamiltonian $H'$ in the accelerating coordinate frame has the same form as H, except for an additional term which can be interpreted as arising from the presence of an effective gravitational field $-\mathbf{a}$. What is then the relation between the momenta $\mathbf{p}_i$ and $\mathbf{p}_i'$ in the two frames?

# CHAPTER VIII

# HAMILTON-JACOBI THEORY

Much of our work to this point has been involved with learning how to write down the equations of motion of a mechanical system in various ways and with studying their general properties. In this chapter we introduce Hamilton-Jacobi theory, which is the most powerful analytic method known for finding the general solution to the mechanical equations of motion.[1] The method involves finding the generating function of a canonical transformation from the original variables to new variables for which the equations of motion are trivial. Apart from its practical aspects, Hamilton-Jacobi theory throws new light on mechanics and its surprisingly intimate connection with optics, a connection which sees its full fruit in wave mechanics.

## Hamilton's principal function

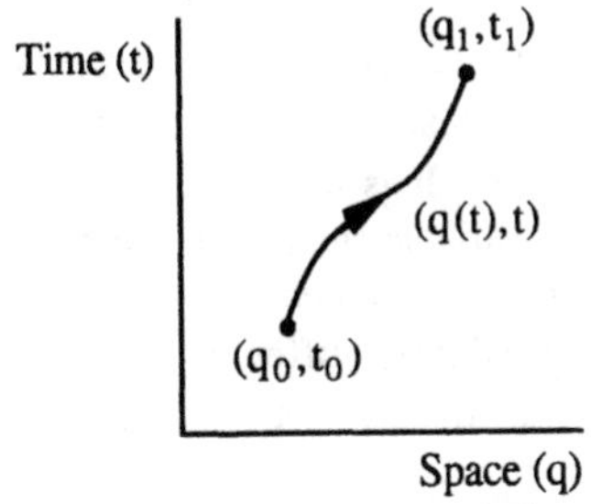


Fig. 8.01. Path in configuration space

The action $S[q(t)]$ for a path $q(t)$ in configuration space (Fig. 8.01), between end points $(q_0,t_0)$ and $(q_1,t_1)$, has been defined as the time integral of the Lagrangian $L(q,\dot{q},t)$ along the path,

$$S[q(t)] = \int_{t_0}^{t_1} L(q,\dot{q},t)\,dt.$$

The actual path which the system follows from $(q_0,t_0)$ to $(q_1,t_1)$ is the one which makes the action stationary. As we have seen in Chapter IV, this implies that the generalized coordinates satisfy Lagrange's equations of motion, a set of f second order differential

---

[1]For parallel reading see: Herbert Goldstein, *Classical Mechanics*, (Addison-Wesley Publishing Company, Reading, 1950; 1980) 2nd ed., Chap. 10; Cornelius Lanczos, *The Variational Principles of Mechanics* (University of Toronto Press, Toronto, 1949; 1962; 1966; 1970) 4th ed.; republished by Dover Publications, Inc., New York, 1986, Chap. VIII.

equations for the coordinates as functions of time. The general solution to these equations has the form

$$q = f(t;c)$$

where the c's are a set of 2f independent assignable constants of integration. These are usually determined by specifying the generalized coordinates q and velocities $\dot{q}$ (or the generalized momenta $p = \partial L/\partial \dot{q}$) at some initial time $t_0$,

$$q_0 = f(t_0;c) \qquad \dot{q}_0 = \frac{df}{dt}(t_0;c).$$

Inverting this set of 2f equations gives the c's in terms of the initial conditions, the $q_0$ and $\dot{q}_0$. Here we do things a little differently. Rather than specifying initial conditions, we specify two points $(q_0,t_0)$ and $(q_1,t_1)$, the ends of the path. We then have

$$q_0 = f(t_0;c) \qquad q_1 = f(t_1;c),$$

which in suitable circumstances[2] we can invert to obtain the c's in terms of $(q_0,t_0)$ and $(q_1,t_1)$. The action along the actual path from $(q_0,t_0)$ to $(q_1,t_1)$ is then a *function* $S(q_1,t_1;q_0,t_0)$ of the end points. It is known as **Hamilton's principal function**. Since we now wish to consider variable end points, we drop the subscript "1" on the final point. In Chapter V we showed that the action $S(q + \Delta q, t + \Delta t; q_0 + \Delta q_0, t_0 + \Delta t_0)$ along a neighboring actual path with slightly different end points $(q_0 + \Delta q_0, t_0 + \Delta t_0)$ and $(q + \Delta q, t + \Delta t)$ differs from $S(q,t;q_0,t_0)$ by an amount

$$\Delta S = (p\Delta q - H\Delta t) - (p_0\Delta q_0 - H_0\Delta t_0).$$

Here $p_0$ and p are the generalized momenta, and $H_0$ and H are the Hamiltonians, at the original end points. We can read off from this the derivatives of $S(q,t;q_0,t_0)$ with respect to its $2f + 2$ arguments. The derivatives with respect to the initial coordinates $q_0$ and the final coordinates q are

$$\frac{\partial S}{\partial q_0} = -p_0 \qquad \frac{\partial S}{\partial q} = p.$$

The first of these equations gives the momentum $p_0$ with which the system must leave $q_0$ at time $t_0$ if it is to arrive at q at time t; the second equation then gives the momentum p with which it arrives. We can invert these equations to find

---

[2]V.I. Arnol'd, *Mathematical Methods of Classical Mechanics*, (Springer-Verlag New York, Inc. 1978), trans. K. Vogtmann and A. Weinstein, section 46.

$$q = f(t;q_0,p_0,t_0) \qquad p = g(t;q_0,p_0,t_0),$$

which express the present time t variables $(q,p)$ in terms of the initial variables $(q_0,p_0)$. We see that Hamilton's principal function $S(q,t;q_0,t_0)$ is the (type 1) generating function of a canonical transformation from old variables $(q,p)$ to new variables $(q_0,p_0)$ (and $-S$ is the generating function of the transformation from $(q_0,p_0)$ to $(q,p)$). This transformation contains both the initial time $t_0$ and the final time t as parameters. Our expression for the differential $\Delta S$ shows that

$$\frac{\partial S}{\partial t_0} = H_0 \qquad \frac{\partial S}{\partial t} = -H.$$

What is the content of these equations? Suppose that $(q,p)$ are canonical variables with dynamics governed by Hamilton's equations with Hamiltonian $H = H(q,p,t)$,

$$\frac{dq}{dt} = \frac{\partial H}{\partial p} \qquad \frac{dp}{dt} = -\frac{\partial H}{\partial q}.$$

Then the new variables $(q_0,p_0)$ are canonical variables with dynamics governed by Hamilton's equations with Hamiltonian $K = H + \dfrac{\partial S}{\partial t} = 0$. The new Hamiltonian is zero, so the dynamics of the new variables is very simple,

$$\frac{dq_0}{dt} = 0 \qquad \frac{dp_0}{dt} = 0;$$

the new variables are constant. This, of course, is what we want: the initial variables $(q_0,p_0)$ are indeed a set of 2f t-fixed independent constants. We can turn this around and say that with $(q_0,p_0)$ fixed, the equations of the canonical transformation generated by S give a solution, $(q,p)$ as functions of time t, to Hamilton's equations with Hamiltonian H, which contains $2f+1$ constants, $(q_0,p_0)$ *and* $t_0$. However, it is not yet clear that $(q_0,p_0)$ are the *initial values* of $(q,p)$. Also, there seems to be one too many constants. The explanation of this is interesting and involves the equation $\partial S/\partial t_0 = H_0$ which we have not yet used. First note that specifying the 2f quantities $(q_0,p_0)$ at some *particular* time $t_0$, for instance at $t_0 = 0$, is all that is required to specify completely a trajectory in configuration or phase space. Now note that all those points $(q_0',p_0',t_0')$ in phase space which lie along the actual path through $(q_0,p_0,t_0 = 0)$ are equivalent "initial conditions"; they yield the same $q(t)$ and $p(t)$. The $2f+1$ constants $(q_0,p_0,t_0)$ (now dropping the primes) are thus not independent. To say that the $(q_0,p_0,t_0)$ lie along the actual path means that $(q_0,p_0)$, regarded as functions of $t_0$ for fixed $(q,p,t)$, satisfy Hamilton's equations with Hamiltonian $H_0 = H(q_0,p_0,t_0)$

$$\frac{dq_0}{dt_0} = \frac{\partial H_0}{\partial p_0} \qquad \frac{dp_0}{dt_0} = -\frac{\partial H_0}{\partial q_0}.$$

Now $-S$ is the generating function of a canonical transformation from $(q_0, p_0)$ to $(q, p)$. This is just the opposite of what we started with and, proceeding as before, we see that $(q, p)$, regarded as functions of $t_0$, satisfy Hamilton's equations with Hamiltonian $H_0 + \partial(-S)/\partial t_0 = 0$. That is, they are constant. Alternatively, if we keep $(q, p)$ "$t_0$-fixed," then the appropriate $(q_0, p_0)$ satisfy, as functions of $t_0$, Hamilton's equations with Hamiltonian $H_0$. They behave as initial values should.

Perhaps an example will help clarify the concepts involved. Let us consider a free particle in one dimension. Such a system goes from $(q_0, t_0)$ to $(q, t)$ at constant velocity $\frac{q - q_0}{t - t_0}$, and thus at constant Lagrangian (= kinetic energy) $\frac{m}{2}\left(\frac{q - q_0}{t - t_0}\right)^2$. Hamilton's principal function is therefore

$$S(q, t; q_0, t_0) = \frac{m}{2}\frac{(q - q_0)^2}{t - t_0}.$$

The equations of the canonical transformation generated by S are

$$p = \frac{\partial S}{\partial q} = m\frac{q - q_0}{t - t_0} \qquad p_0 = -\frac{\partial S}{\partial q} = +m\frac{q - q_0}{t - t_0},$$

which yield

$$q = q_0 + (p_0/m)(t - t_0) \qquad p = p_0.$$

Since $\frac{\partial S}{\partial t} = -\frac{m}{2}\left(\frac{q - q_0}{t - t_0}\right)^2 = -H$ (= −kinetic energy), the transformed Hamiltonian is zero, and $(q_0, p_0)$, as functions of t, are constant. The equations of the canonical transformation then give the solution, $(q, p)$ as functions of t, to the dynamical problem. We further note that all sets $(q_0, p_0, t_0)$ for which $q_0 - (p_0/m)t_0$ and $p_0$ are fixed are equivalent, yield the same $(q(t), p(t))$. These are the sets for which $(q_0, p_0)$, regarded as functions of $t_0$, lie on the actual path.

## Jacobi's complete integral

We have seen that Hamilton's principal function $S(q,t;q_0,t_0)$ is the generating function of the canonical transformation from the variables $(q,p)$ at time t to the initial variables $(q_0,p_0)$ at time $t_0$. S contains the fully integrated solution to the dynamical problem; all that is left to do are differentiations and eliminations. We have also seen how to find S once the solution to the dynamical problem is known. We now wish to reverse the procedure: to devise some independent way first to find S, and then use it to write down the solution to the dynamical problem. Following Jacobi we first consider the less particular problem of finding a (time-dependent) generating function of a canonical transformation from variables $(q,p)$ at time t to new variables $(\beta,\alpha)$ *which are constant in time*. We do not specify these new variables further. We can take the generating function to be a function of the old coordinates q (the present, time t, coordinates) and the new momenta $\alpha$, which gives a type 2 generating function $S(q;\alpha;t)$. We denote this generating function by the same letter "S" as we used for Hamilton's principal function; if it is necessary to distinguish between them, we add a subscript H = Hamilton or J = Jacobi. The equations of the canonical transformation read

$$\beta = \frac{\partial S(q;\alpha;t)}{\partial \alpha} \qquad p = \frac{\partial S(q;\alpha;t)}{\partial q}.$$

The first set of f equations gives the new (constant) coordinates $\beta$. They can be inverted to obtain the coordinates q as functions of time t and the 2f constants $(\beta,\alpha)$. Substituting the result into the second set of f equations then gives the momenta p as functions of time t and the constants $(\beta,\alpha)$. We thus obtain

$$q = f(t;\beta,\alpha) \qquad p = g(t;\beta,\alpha).$$

These equations, which give the canonical variables as functions of time and which contain 2f independent assignable constants $(\beta,\alpha)$, are the general solution to the dynamical problem. Note that this procedure automatically produces the correct number of constants. If we wish, we can express the constants $(\beta,\alpha)$ in terms of the initial (time $t_0$) values of the canonical variables $(q_0,p_0)$ by setting

$$q_0 = f\left(t_0;\beta,\alpha\right) \qquad p_0 = g\left(t_0;\beta,\alpha\right),$$

and then inverting these 2f equations. However, we do not have to do this; constants other than the initial values are often more convenient.

We have not yet considered how to find $S(q;\alpha;t)$. Since S generates a canonical transformation to new variables which are constant in time, the new Hamiltonian $K = H + \partial S/\partial t$ satisfies $\partial K/\partial\alpha = 0$, $\partial K/\partial\beta = 0$ and is thus independent of the new variables. It is at most a function of time, and we can without loss take it to be zero. We then have

$$H\left(q,\frac{\partial S}{\partial q},t\right)+\frac{\partial S}{\partial t}=0,$$

where we have also used the equations of the canonical transformation to replace the momenta p in H by $\partial S/\partial q$. This is the famous **Hamilton-Jacobi equation**. It is a first order partial differential equation for $S(q;\alpha;t)$ with $f+1$ independent variables, the f coordinates q and the time t. It is perhaps not immediately obvious where the new momenta $\alpha$ in $S(q;\alpha;t)$ come from. We must find what is known as a **complete integral** of the Hamilton-Jacobi partial differential equation. By definition, a complete integral contains as many independent assignable constants as there are independent variables, in this case $f+1$ of them, $\alpha_1,\alpha_2,...,\alpha_f,\alpha_{f+1}$. Since only the space and time derivatives of S, and not S itself, appear in the Hamilton-Jacobi equation, any solution S can be modified by adding to it an arbitrary constant. We can take one of the constants $\alpha$, say $\alpha_{f+1}$, to be this purely additive constant of integration. This constant has no effect on the transformation, so we can ignore it, writing the complete integral in the form

$$S(q_1,q_2,...,q_f;\alpha_1,\alpha_2,...,\alpha_f;t)$$

where none of the constants $\alpha$ is purely additive.[3] These constants $\alpha$ are the new momenta. The function S is known as **Jacobi's complete integral**.

If we are interested simply in writing down a general solution to a dynamical problem, Jacobi's complete integral is all we require; there is no need to find Hamilton's principal function. It is nevertheless of interest to consider further the connection between Jacobi's complete integral $S_J(q,\alpha,t)$ and Hamilton's principal function $S_H(q,t;q_0,t_0)$. As we have seen, $S_H$ is the generating function of a canonical transformation from present (time t) variables $(q,p)$ to initial (time $t_0$) variables $(q_0,p_0)$, whereas $S_J$ is the generating function of a canonical transformation from present (time t) variables $(q,p)$ to variables $(\beta,\alpha)$ which are constant. We can accomplish the canonical transformation generated by $S_H$ in two steps by using $S_J$: transform first from $(q,p)$ at time t to $(\beta,\alpha)$ using $S_J(q,\alpha,t)$, and then from $(\beta,\alpha)$ to $(q_0,p_0)$ at time $t_0$ using $-S_J(q_0,\alpha,t_0)$ (Fig. 8.02). Since the generating function of this two-step process is the sum of the generating functions of the steps, we have

$$S_H(q,t;q_0,t_0)=S_J(q;\alpha;t)-S_J(q_0;\alpha;t_0).$$

[3]We also require $\det|\partial^2 S/\partial q_a \partial\alpha_b|\neq 0$, so that the equations of the canonical transformation generated by S can be inverted.

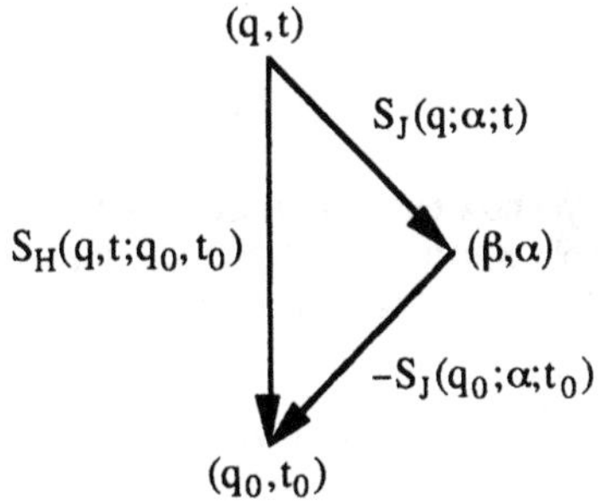


Fig. 8.02. $S_H$ in terms of $S_J$

The intermediate point $(\beta,\alpha)$ we end up on for the first step and start off from for the second step must of course be the same point. This means that the $S_J$'s on the right are the *same* complete integrals, and that

$$\beta = \frac{\partial S_J(q;\alpha;t)}{\partial \alpha} = \frac{\partial S_J(q_0;\alpha;t_0)}{\partial \alpha}.$$

That is, the $\alpha$'s in the above $S_H$ must be such that

$$\frac{\partial S_H}{\partial \alpha} = 0.$$

This set of f equations gives the $\alpha$'s in terms of $(q,t,q_0,t_0)$. As we may easily check and as required, $S_H$ satisfies the equations

$$\frac{\partial S_H(q,t;q_0,t_0)}{\partial q_0} = -\frac{\partial S_J(q_0;\alpha;t_0)}{\partial q_0} = -p_0 \qquad \frac{\partial S_H(q,t;q_0,t_0)}{\partial q} = \frac{\partial S_J(q;\alpha;t)}{\partial q} = p$$
$$\frac{\partial S_H(q,t;q_0,t_0)}{\partial t_0} = -\frac{\partial S_J(q_0;\alpha;t_0)}{\partial t_0} = H_0 \qquad \frac{\partial S_H(q,t;q_0,t_0)}{\partial t} = \frac{\partial S_J(q;\alpha;t)}{\partial t} = -H\ .$$

The partial derivatives are to be performed keeping the other indicated variables fixed. Note that although the $\alpha$'s in the $S_J$'s are functions of $(q,t,q_0,t_0)$, this dependence need not be considered since $\partial S_H/\partial \alpha = 0$.

As an example of Jacobi's approach, let us consider again a free particle in one dimension. The Hamilton-Jacobi equation is

$$\frac{1}{2m}\left(\frac{\partial S}{\partial q}\right)^2 + \frac{\partial S}{\partial t} = 0,$$

with a complete integral

$$S(q,E,t) = \sqrt{2mE}\,q - Et.$$

While it may not be obvious yet how this is obtained, it is clearly a complete integral: it is a solution to the Hamilton-Jacobi equation which contains the one non-additive constant E.[4] The equations of the canonical transformation give

$$\beta_E = \frac{\partial S}{\partial E} = \sqrt{\frac{m}{2E}}\,q - t \quad \text{which yields} \quad q = \sqrt{\frac{2E}{m}}\,(t + \beta_E)$$

and

$$p = \frac{\partial S}{\partial q} = \sqrt{2mE}.$$

This is the general solution to the dynamical equations, giving $(q,p)$ as functions of time and containing two constants $(\beta_E, E)$; E, the energy, is the "new momentum" and $\beta_E$ is the "new coordinate" conjugate to E.

Now let's see how we can recover Hamilton's principal function from the above Jacobi complete integral. We set

$$\begin{aligned} S_H(q,t;q_0,t_0) &= S_J(q,E,t) - S_J(q_0,E,t_0) \\ &= \sqrt{2mE}\,(q - q_0) - E(t - t_0) \end{aligned}$$

and determine E by requiring that

$$\frac{\partial S_H}{\partial E} = \sqrt{\frac{m}{2E}}\,(q - q_0) - (t - t_0) = 0.$$

This gives $E = \frac{m}{2}\left(\frac{q - q_0}{t - t_0}\right)^2$, and substituting back we find

$$S_H(q,t;q_0,t_0) = \frac{m}{2}\frac{(q - q_0)^2}{t - t_0}$$

in agreement with our earlier calculation.

---

[4]E is positive; the square root can be positive or negative.

## Time-independent Hamilton-Jacobi equation

Quite often the Hamiltonian does not depend explicitly on time. The Hamilton-Jacobi equation then becomes

$$H\left(q,\frac{\partial S}{\partial q}\right)+\frac{\partial S}{\partial t}=0.$$

In these cases the time dependence can be eliminated once and for all. We try a solution of the form

$$S(q,t)=W(q)+T(t),$$

the sum of a function W of the coordinates q alone and a function T of the time t alone. Substituting this into the Hamilton-Jacobi equation , we find

$$H\left(q,\frac{\partial W}{\partial q}\right)=-\frac{dT}{dt}.$$

The left-hand side is independent of t, whereas the right-hand side is independent of the coordinates q. Both sides must therefore equal a constant which we denote by E,

$$H\left(q,\frac{\partial W}{\partial q}\right)=E \qquad \frac{dT}{dt}=-E.$$

We call the first of these equations the **time-independent Hamilton-Jacobi equation**. A Jacobi complete integral W of this equation is a function of the f coordinates q and (including E) of f non-additive independent constants $\alpha$. E is the constant value of the Hamiltonian, which in common cases is the total energy. It is sometimes convenient to take E as one of the f constants $\alpha$. In other cases a set $(\alpha_1,\alpha_2,\ldots,\alpha_f)$ not including E is more convenient; E is then some function of these, $E=E(\alpha_1,\alpha_2,\ldots,\alpha_f)$. The equation for the time dependence can be integrated trivially, so a complete integral to the Hamilton-Jacobi equation for time-independent Hamiltonians has the form

$$S=W(q_1,q_2,\ldots,q_f;\alpha_1,\alpha_2,\ldots,\alpha_f)-E(\alpha_1,\alpha_2,\ldots,\alpha_f)t.$$

The equations of the canonical transformation then give

$$p=\frac{\partial W(q;\alpha)}{\partial q} \qquad \beta+\frac{\partial E(\alpha)}{\partial \alpha}t=\frac{\partial W(q;\alpha)}{\partial \alpha}.$$

These equations show that Jacobi's complete integral W is itself the generating function of a (time-independent) canonical transformation, from a set of old variables $(q,p)$ to a set of new variables

$$Q = \beta + \frac{\partial E}{\partial \alpha} t \qquad P = \alpha$$

for which the new coordinates Q are all cyclic. The new Hamiltonian for this set of variables is simply E(P).

If we take E as one of the new momenta, say $E = \alpha_1$, then the second half of the above transformation equations becomes

$$\beta_1 + t = \frac{\partial W(q;E,\alpha_2,\cdots,\alpha_f)}{\partial E}$$

$$\beta_a = \frac{\partial W(q;E,\alpha_2,\cdots,\alpha_f)}{\partial \alpha_a} \qquad \text{where} \quad a = 2,\cdots,f\ .$$

The shape of the trajectory in configuration space is given by the last $f-1$ time-independent equations, and the motion along the trajectory in time is given by the first equation.

For example, the time-independent Hamilton-Jacobi equation for a particle of mass m moving in one dimension in a potential $V(x)$ is

$$\frac{1}{2m}\left(\frac{dW}{dx}\right)^2 + V(x) = E,$$

with solution

$$W = \int \sqrt{2m(E - V(x))}\,dx.$$

It is simplest to take E itself as the new momentum $\alpha$. The transformation equations then give the solution to the dynamical problem,

$$p = \frac{\partial W}{\partial x} = \sqrt{2m(E - V(x))} \qquad \beta + t = \frac{\partial W}{\partial E} = \sqrt{\frac{m}{2}} \int \frac{dx}{\sqrt{2m(E - V(x))}}.$$

In particular, for a simple harmonic oscillator with $V = \frac{1}{2} m\omega^2 x^2$ we have

$$W = \int \sqrt{2m\left(E - \tfrac{1}{2} m\omega^2 x^2\right)}\,dx.$$

The integration can be performed by setting

$$x = \sqrt{\frac{2E}{m\omega^2}} \sin\phi \qquad dx = \sqrt{\frac{2E}{m\omega^2}} \cos\phi\, d\phi,$$

so that

$$W = (2E/\omega)\int \cos^2\phi\, d\phi = (E/\omega)(\phi + \sin\phi\cos\phi).$$

This is shown as a function of x in Fig. 8.03. Note that as x is taken through one cycle, the phase $\phi$ increases by $2\pi$, and W increases by $2\pi E/\omega$. Continuity of W in x can be maintained by adding to the above expression, if necessary,[5] an appropriate constant for each cycle. These constants do not affect the role of W as a generating function, but the resulting W is a multivalued function of x.

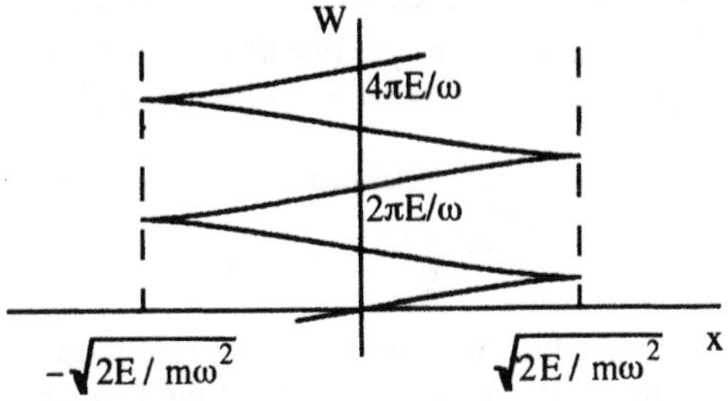


Fig. 8.03. W for a simple harmonic oscillator

The solution to the dynamical problem is obtained by differentiating this expression for W with respect to x and E (remember that $\phi$ is a function of x and E), or more simply by first differentiating the integral expression for W given above and then integrating. Either way we find

$$p = \frac{\partial W}{\partial x} = \sqrt{2mE}\cos\phi \qquad \text{with} \qquad x = \sqrt{\frac{2E}{m\omega^2}}\sin\phi$$

and

$$\beta + t = \frac{\partial W}{\partial E} = \frac{\phi}{\omega},$$

which is the well-known solution to the simple harmonic oscillator problem.

The generating functions we have been considering so far in this section have been of "Jacobi type." To reflect this, we now write them $S_J$ and $W_J$. It is of interest to consider also the "Hamilton type" functions $S_H$ and $W_H$ for time-independent Hamiltonians. Using the previously established connection between the two types, we have for Hamilton's principal function

$$S_H(q,t;q_0,t_0) = W_H(q,q_0,E) - E(t - t_0)$$

where

$$W_H(q,q_0,E) = W_J(q,E,\alpha) - W_J(q_0,E,\alpha).$$

[5]This depends on the branch we use in determining $\phi$ from x.

The $f-1$ constants $(\alpha_2,\cdots,\alpha_f)$ on the right are to be expressed in terms of $(q,q_0,E)$ by using the $f-1$ equations

$$\frac{\partial W_H}{\partial \alpha_a}=0 \quad \text{where} \quad a=2,\cdots,f.$$

Note that $W_J$ contains f constants, the $f-1$ $\alpha$'s and E, whereas $W_H$ contains $f+1$ constants, the f $q_0$'s and E. The consequences of this will become apparent shortly. To complete the passage to Hamilton's principal function $S_H$, we should also express the energy E in terms of $(q,t;q_0,t_0)$ by using the equation $\partial S_H/\partial E=0$, which becomes in this case

$$\frac{\partial W_H(q,q_0,E)}{\partial E}=t-t_0 .$$

We see that for time-independent Hamiltonians, Hamilton's principal function $S_H(q,t;q_0,t_0)$ depends only on the difference $t-t_0$ of the final and initial times, as we might expect. The function $W_H(q,q_0,E)$ is of interest in its own right. It is called **Hamilton's characteristic function**. Its derivatives with respect to the final and initial coordinates are given by

$$\frac{\partial W_H(q,q_0,E)}{\partial q}=\frac{\partial W_J(q,E,\alpha)}{\partial q}=p$$
$$\frac{\partial W_H(q,q_0,E)}{\partial q_0}=-\frac{\partial W_J(q_0,E,\alpha)}{\partial q_0}=-p_0 .$$

The $\alpha$'s in the middle expressions here are functions of $(q,q_0,E)$, but they are to be held fixed in the differentiations with respect to q and $q_0$. At first sight these look like the usual equations of a canonical transformation from $(q,p)$ to $(q_0,p_0)$ with type 1 generating function $W_H(q,q_0,E)$ (or from $(q_0,p_0)$ to $(q,p)$ with generating function $-W_H(q,q_0,E)$). But this clearly cannot be: we know that specifying a final point $(q,p)$ does not fix a single initial point $(q_0,p_0)$ but rather a one-dimensional infinity, a line, of initial points all of which lead to $(q,p)$. The difficulty with $W_H$ as a generating function comes when we try to invert the first set of equations to obtain the $q_0$'s as functions of the $(q,p)$. We can't do it; the transformation is singular with $\det|\partial^2 W_H/\partial q_0 \partial q|=0$. We can see this most clearly from the middle expressions above. The f $q_0$'s are contained in the $f-1$ $\alpha$'s, so one of the $q_0$'s must remain undetermined. Letting this $q_0$ take on all possible values then gives the line of initial points $(q_0,p_0)$ for the given final point $(q,p)$, as required.

## Separation of variables

For systems with more than one degree of freedom we attempt to integrate the Hamilton-Jacobi equation by using the **method of separation of variables**. Whether this method works or not depends on the system being considered and on the choice of generalized coordinates. If it does work, the method provides a powerful yet straightforward approach to the solution of a dynamical problem. We have already used this approach to separate out the time for systems with time-independent Hamiltonians. Basically the method consists of trying a solution of the form

$$W(q_1,\cdots,q_f;\alpha_1,\cdots,\alpha_f)=\sum_{a=1}^{f}W_a(q_a;\alpha_1,\cdots,\alpha_f),$$

a sum of f terms each depending on a different one of the generalized coordinates. We then try to rearrange the Hamilton-Jacobi equation so that one side of the equation depends only on one of the coordinates, $q_1$ for instance, while the other side depends only on the other coordinates $(q_2,\cdots,q_f)$. Both sides of the equation can then be set equal a **separation constant**. We try to continue the process until the variables are completely separated. If we succeed, each of the resulting f equations then yields the corresponding $W_a$, and we also pick up f non-additive separation constants. That is, we obtain a complete integral. We illustrate the method with some examples.[6]

## Free particle, in cartesian coordinates

The Hamiltonian is

$$H=\frac{1}{2m}(p_x^2+p_y^2+p_z^2),$$

and hence the time-independent Hamilton-Jacobi equation is

$$\frac{1}{2m}\left[\left(\frac{\partial W}{\partial x}\right)^2+\left(\frac{\partial W}{\partial y}\right)^2+\left(\frac{\partial W}{\partial z}\right)^2\right]=E.$$

We try a solution of the form

$$W=X(x)+Y(y)+Z(z).$$

The Hamilton-Jacobi equation becomes

[6]Others may be found in L. D. Landau and E. M. Lifshitz, *Mechanics*, (Pergamon Press, Oxford, 1960; 1969; 1976), 3rd ed., trans. J. B. Sykes and J. S. Bell, pp. 151-153.

$$\frac{1}{2m}\left[\left(\frac{dX}{dx}\right)^2+\left(\frac{dY}{dy}\right)^2+\left(\frac{dZ}{dz}\right)^2\right]=E.$$

The first term on the left-hand side is a function only of x, the second a function only of y, and the third a function only of z. Each must equal a constant. We set

$$\left(\frac{dX}{dx}\right)^2=\alpha_x^2 \qquad \left(\frac{dY}{dy}\right)^2=\alpha_y^2 \qquad \left(\frac{dZ}{dz}\right)^2=\alpha_z^2$$

where $\alpha_x$, $\alpha_y$, $\alpha_z$ are constants. In introducing these we have set $(dX/dx)^2$ equal $\alpha_x^2$ rather than $\alpha_x$, for example, merely to reflect the fact that we know this constant to be positive. The energy E is related to these α's by

$$E=\frac{1}{2m}(\alpha_x^2+\alpha_y^2+\alpha_z^2).$$

Integration of the equations for X, Y, and Z gives

$$W=\alpha_x x+\alpha_y y+\alpha_z z,$$

and the transformation equations then give the solution to the dynamical problem,

$$p_x=\alpha_x \qquad \beta_x+\frac{\alpha_x}{m}t=x$$

$$p_y=\alpha_y \qquad \beta_y+\frac{\alpha_y}{m}t=y$$

$$p_z=\alpha_z \qquad \beta_z+\frac{\alpha_z}{m}t=z\,.$$

We recognize $(\alpha_x,\alpha_y,\alpha_z)$ as the cartesian components of the (constant) linear momentum, and $(\beta_x,\beta_y,\beta_z)$ as the initial ($t=0$) values of the cartesian coordinates.

## Central force, in spherical polar coordinates

The Hamiltonian is

$$H = \frac{1}{2m}\left(p_r^2 + \frac{1}{r^2}p_\theta^2 + \frac{1}{r^2\sin^2\theta}p_\phi^2\right) + V(r) = E,$$

and the time-independent Hamilton-Jacobi equation is

$$\frac{1}{2m}\left[\left(\frac{\partial W}{\partial r}\right)^2 + \frac{1}{r^2}\left(\frac{\partial W}{\partial \theta}\right)^2 + \frac{1}{r^2\sin^2\theta}\left(\frac{\partial W}{\partial \phi}\right)^2\right] + V(r) = E.$$

We try a solution of the form

$$W = R(r) + \Theta(\theta) + \Phi(\phi).$$

The Hamilton-Jacobi equation becomes

$$\frac{1}{2m}\left[\left(\frac{dR}{dr}\right)^2 + \frac{1}{r^2}\left(\frac{d\Theta}{d\theta}\right)^2 + \frac{1}{r^2\sin^2\theta}\left(\frac{d\Phi}{d\phi}\right)^2\right] + V(r) = E,$$

which can be rearranged in the form

$$2mr^2\sin^2\theta\left\{\frac{1}{2m}\left[\left(\frac{dR}{dr}\right)^2 + \frac{1}{r^2}\left(\frac{d\Theta}{d\theta}\right)^2\right] + V(r) - E\right\} = -\left(\frac{d\Phi}{d\phi}\right)^2$$

The left-hand side is independent of $\phi$, whereas the right-hand side is independent of r and $\theta$. Both sides must equal a (negative) constant, which we set equal $-L_z^2$. We shall see later that $L_z$ is the (constant) z-component of the angular momentum. We then have

$$\frac{1}{2m}\left[\left(\frac{dR}{dr}\right)^2 + \frac{1}{r^2}\left(\frac{d\Theta}{d\theta}\right)^2\right] + V(r) + \frac{L_z^2}{2mr^2\sin^2\theta} = E \qquad \left(\frac{d\Phi}{d\phi}\right)^2 = L_z^2.$$

The equation in r and $\theta$ can be rearranged in the form

$$2mr^2\left[\frac{1}{2m}\left(\frac{dR}{dr}\right)^2 + V(r) - E\right] = -\left[\left(\frac{d\Theta}{d\theta}\right)^2 + \frac{L_z^2}{\sin^2\theta}\right].$$

The left-hand side is independent of $\theta$ and the right-hand side is independent of r, so both must equal a (negative) constant, which we set equal $-L^2$. We shall see later that L is the magnitude of the total angular momentum. We have

$$\frac{1}{2m}\left(\frac{dR}{dr}\right)^2 + V(r) + \frac{L^2}{2mr^2} = E \qquad \left(\frac{d\Theta}{d\theta}\right)^2 + \frac{L_z^2}{\sin^2\theta} = L^2.$$

The variables are now completely separated, and we can rearrange and integrate to obtain

$$R(r) = \sqrt{2m}\int\sqrt{E - V(r) - \frac{L^2}{2mr^2}}\,dr$$

$$\Theta(\theta) = \int\sqrt{L^2 - \frac{L_z^2}{\sin^2\theta}}\,d\theta \qquad \Phi(\phi) = L_z\phi$$

and hence find

$$W = \sqrt{2m}\int\sqrt{E - V(r) - \frac{L^2}{2mr^2}}\,dr + \int\sqrt{L^2 - \frac{L_z^2}{\sin^2\theta}}\,d\theta + L_z\phi.$$

The Jacobi complete integral W is the generating function of a canonical transformation from $(r,\theta,\phi;p_r,p_\theta,p_\phi)$ to new coordinates which are cyclic and new momenta which are constant. We can take these new momenta to be the separation constants E, L, $L_z$. The first half of the transformation equations gives

$$p_r = \frac{\partial W}{\partial r} = \sqrt{2m}\sqrt{E - V(r) - \frac{L^2}{2mr^2}}$$

$$p_\theta = \frac{\partial W}{\partial \theta} = \sqrt{L^2 - \frac{L_z^2}{\sin^2\theta}} \qquad p_\phi = \frac{\partial W}{\partial \phi} = L_z\ .$$

The signs of the square roots in $p_r = m\dot{r}$ and $p_\theta = mr^2\dot{\theta}$ (and in W) must be chosen so that they are positive where r and $\theta$ are increasing and negative where r and $\theta$ are decreasing. We can now recognize that $L_z = p_\phi = mr^2\sin^2\theta\,\dot{\phi}$ is the z-component of the angular momentum, and that $L^2 = p_\theta^2 + (p_\phi/\sin\theta)^2 = m^2r^4(\dot{\theta}^2 + \sin^2\theta\,\dot{\phi}^2)$ is the square of the total angular momentum. This identifies and justifies our names for the constants $L_z$ and L.

Now consider the second half of the transformation equations, which involves the derivatives of W with respect to the new momenta E, L, $L_z$. We can proceed in either of two ways: perform the integrations first to find W and then differentiate, or differentiate and then integrate. Since we are more interested here in the equations for the orbit than in W itself, we adopt the second alternative. This gives

$$\beta_E + t = \frac{\partial W}{\partial E} = \sqrt{\frac{m}{2}} \int \frac{dr}{\sqrt{E - V(r) - \dfrac{L^2}{2mr^2}}}$$

$$\beta_L = \frac{\partial W}{\partial L} = \sqrt{2m} \int \frac{dr}{\sqrt{E - V(r) - \dfrac{L^2}{2mr^2}}} \left(\frac{-L}{2mr^2}\right) + \int \frac{L\,d\theta}{\sqrt{L^2 - \dfrac{L_z^2}{\sin^2\theta}}}$$

$$\beta_{L_z} = \frac{\partial W}{\partial L_z} = \int \frac{d\theta}{\sqrt{L^2 - \dfrac{L_z^2}{\sin^2\theta}}} \left(\frac{-L_z}{\sin^2\theta}\right) + \phi\ .$$

The second of these equations gives a functional relation between r and $\theta$, and the third gives a relation between $\theta$ and $\phi$; together they give the shape of the orbit. The first equation gives a functional relation between r and t; it describes how the particle moves along the orbit in time.

We must now do the integrations. We begin with the $\theta$-integration in the third equation. Let $L_z = L\cos i$ where i $(0 \le i \le \pi)$ is the angle between the angular momentum vector and the polar (z) axis. We then have

$$\phi - \beta_{L_z} = \int \frac{\cos i\, d\theta}{\sin^2\theta\sqrt{1 - \dfrac{\cos^2 i}{\sin^2\theta}}} = \int \frac{\cot i\, d\theta}{\sin^2\theta\sqrt{1 - \cot^2 i\cot^2\theta}}$$

$$= \overline{\phi} \qquad \text{on setting } \cot i\cot\theta = \sin\overline{\phi} \text{ and } \frac{\cot i\, d\theta}{\sin^2\theta} = -\cos\overline{\phi}\, d\overline{\phi}\ .$$

To sort out the sign, note that, for $0 \le i \le \pi/2$, $\theta$ oscillates back and forth between $\pi/2 + i$ and $\pi/2 - i$.[7] These are the values which make the square root zero and are where $d\theta$ and the square root change sign. From $\theta = \pi/2 + i$ through $\pi/2$ to $\pi/2 - i$, $d\theta$ is negative. For this half cycle and for $0 \le i \le \pi/2$,[8] the variable $\overline{\phi}$ increases from $-\pi/2$ through 0 to $\pi/2$. We must thus take $\sqrt{\cos^2\overline{\phi}} = -\cos\overline{\phi}$ so that the square root for this half cycle is negative as well. The end result of the integration is

[7]For $\pi/2 < i \le \pi$, $\theta$ oscillates between $\pi/2 + (\pi - i)$ and $\pi/2 - (\pi - i)$.

[8]For $\pi/2 < i \le \pi$, $\overline{\phi}$ decreases from $\pi/2$ through 0 to $-\pi/2$.

$$\cot i \cot\theta = \sin(\phi - \beta_{L_z}).$$

This can be written

$$\sin i \sin\theta \sin(\phi - \beta_{L_z}) = \cos i \cos\theta$$

and can be identified as the equation of a plane passing through the origin.

To see this and to connect with standard astronomical terminology as used in the description of planetary orbits,[9] we set up a system of cartesian coordinates with the origin at the sun, the x- and y-axes in the plane of the ecliptic with the x-axis pointing towards the vernal equinox, and the z-axis perpendicular to the plane of the ecliptic and pointing "north" (Fig. 8.04). The orbit of a planet lies in a plane, the **orbital plane**, passing through the sun. The orientation of this plane is described first by giving the angle i through which the orbital plane is tipped with respect to the plane of the ecliptic. This angle i is the **inclination** of the orbit. The intersection of the orbital plane and the plane of the ecliptic is the **line of nodes**. Where the planetary orbit passes through the plane of the ecliptic on its way from "south" to "north" is the **ascending node**, and where it passes through the plane of the ecliptic on its way from "north" to "south" is the **descending node**. We complete the description of the orbital plane by giving the angle $\Omega$ measured "easterly" in the plane of the ecliptic between the x-axis (direction of vernal equinox) and the direction of the ascending node. The angle $\Omega$ is the **longitude of the ascending node**.

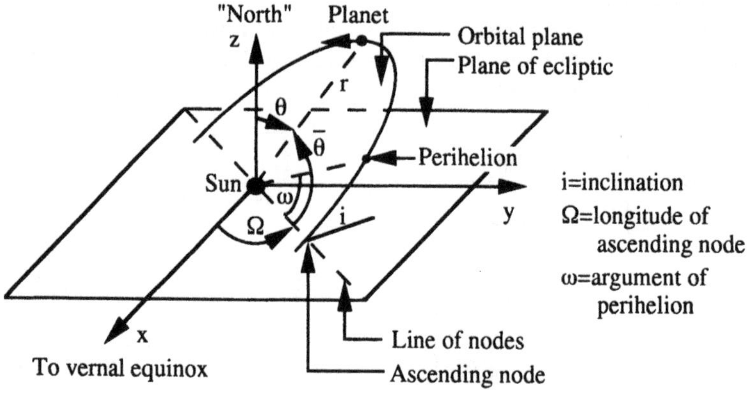


Fig. 8.04. Parameters of planetary orbit

The normal **n** to the orbital plane points in a direction $\theta = i$ and $\phi = \Omega - \pi/2$, and its cartesian components are given by

$$\mathbf{n} = (\sin i \sin\Omega, -\sin i \cos\Omega, \cos i).$$

---

[9]See, for example, Forest Ray Moulton, *An Introduction to Celestial Mechanics*, (Macmillan, New York, NY, 1902, 1914, 1923), 2nd ed..

If we let

$$\mathbf{r} = r(\sin\theta\cos\phi, \sin\theta\sin\phi, \cos\theta)$$

be the radius vector from the sun to the planet, we can write the equation of the orbital plane in the form

$$\mathbf{n}\cdot\mathbf{r} = 0.$$

This gives

$$\sin i \sin\theta \sin(\phi - \Omega) = \cos i \cos\theta,$$

which is what we found previously, with i now identified as the inclination and the generalized coordinate $\beta_{L_z}$ conjugate to $L_z$ identified as the longitude $\Omega$ of the ascending node.

We now look at the $\theta$-integration in the second equation. Again setting $L_z = L\cos i$, we have

$$\int \frac{d\theta}{\sqrt{1 - \dfrac{\cos^2 i}{\sin^2\theta}}} = \int \frac{\sin\theta\, d\theta}{\sqrt{\sin^2 i - \cos^2\theta}} = \bar{\theta}$$

on setting $\cos\theta = \sin i \sin\bar{\theta}$ and $\sin\theta\, d\theta = -\sin i \cos\bar{\theta}\, d\bar{\theta}$ .

The sign can be sorted out as before and requires that $\sqrt{\sin^2 i \cos^2\bar{\theta}} = -\sin i \cos\bar{\theta}$. $\bar{\theta}$ is the angle, measured in the plane of the orbit, between the radius vector and the direction of the ascending node. To check, see Fig. 8.05 (in which the axes have been rotated about the z-axis so that the x'-axis is in the direction of the ascending node), noting that the z-component of a point on the orbit is $r\cos\theta = r\sin\bar{\theta}\sin i$.

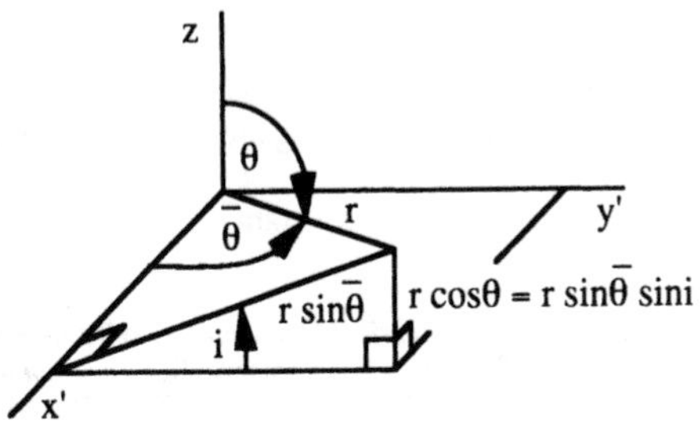


Fig. 8.05. Angles $\theta$ and $\bar{\theta}$

The second equation becomes

$$\overline{\theta} - \beta_L = \frac{L}{\sqrt{2m}} \int \frac{dr}{r^2\sqrt{E - V(r) - \dfrac{L^2}{2mr^2}}},$$

and so we return again to the equation of the orbit first found in Chapter I. As shown there, for the gravitational potential $V = -k/r$ the r-integration gives

$$\frac{a(1-e^2)}{r} = 1 + e\cos(\overline{\theta} - \beta_L),$$

which is the equation of an ellipse (for $E < 0$) of **semi-major axis** $a = k/2|E|$ and **eccentricity** $e = \sqrt{1 - \dfrac{2L^2|E|}{mk^2}}$. The generalized coordinate $\beta_L$ conjugate to the total angular momentum L is the angle measured in the orbital plane from the ascending node to perihelion. In astronomy this angle is called the **argument of perihelion** and is denoted by $\omega$. Finally, the equation for how the planet moves along the orbit in time can be integrated as in Chapter I, showing that the generalized coordinate $\beta_E$ conjugate to E is (minus) the **time of passage of perihelion** $t_0$.

## Hamilton-Jacobi mechanics, geometric optics, and wave mechanics

In this section we explore some of the remarkable connections which exist between mechanics, geometric optics, and wave mechanics.[10] For simplicity we consider a single particle of mass m moving in a potential $V(\mathbf{r})$. The time-independent Hamilton-Jacobi equation is then

$$\frac{1}{2m}|\nabla W|^2 + V(\mathbf{r}) = E$$

where E is the energy of the particle. Suppose that we have a solution $W(\mathbf{r})$ to this equation. We can picture this solution by drawing the family of surfaces (Fig. 8.06)

$$W(\mathbf{r}) = \text{constant}.$$

[10]For parallel reading see: Herbert Goldstein, *Classical Mechanics*, (Addison-Wesley Publishing Company, Reading, 1950; 1980) 2nd ed., pp. 484-492; Cornelius Lanczos, *The Variational Principles of Mechanics* (University of Toronto Press, Toronto, 1949; 1962; 1966; 1970) 4th ed.; republished by Dover Publications, Inc., New York, 1986, pp. 264-280; Max Born and Emil Wolf, *Principles of Optics*, (Pergamon Press, Oxford, 1959; 1964; 1965; 1970; 1975) 5th ed., Chaps. III and IV, and appendices I and II; E. Schrödinger, *Collected Papers on Wave Mechanics* (Chelsea Publishing Company, New York, 1978), 2nd ed.. Trans. of *Abhandlungen zur Wellenmechanik* (Johann Ambrosius Barth, 1928) by J. F. Shearer and W. M. Deans., pp. 13-30.

This is analogous to the situation in electrostatics, where we picture the electrostatic potential by drawing the family of equipotential surfaces. At any point **r**, the momentum **p** of the particle is given by the gradient of W,

$$\mathbf{p}(\mathbf{r}) = \nabla W(\mathbf{r}).$$

Its direction is perpendicular to the local surface $W(\mathbf{r}) = \text{constant}$ in the direction of increasing W, and its magnitude is given by

$$p(\mathbf{r}) = |\nabla W(\mathbf{r})| = \sqrt{2m(E - V(\mathbf{r}))},$$

the second equality following from the Hamilton-Jacobi equation. Momentum is analogous to (minus) the electric field. Just as we can go from W to **p** by differentiation, we can go from **p** to W by integration, the difference in W between any two points being the line integral of **p** along a curve joining the points,

$$\Delta W = \int \mathbf{p}(\mathbf{r}) \cdot d\mathbf{r}.$$

Since **p** is a gradient, all curves joining these two given points which can be deformed continuously into one another give the same $\Delta W$. However, there are usually sets of non-equivalent curves, and as a result $\Delta W$ is usually multivalued; see, for example, Fig. 8.03.

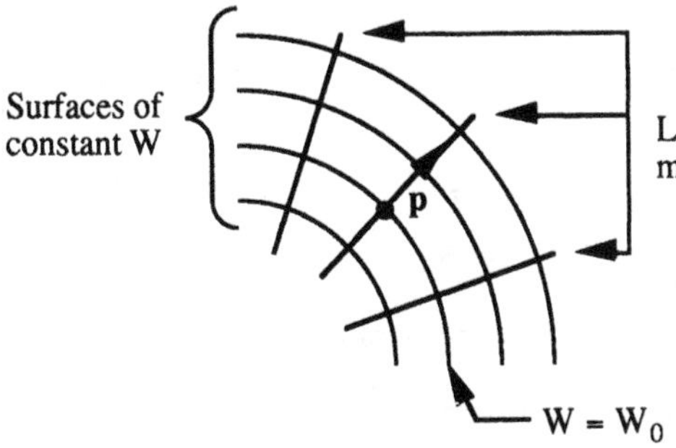

Fig. 8.06. Surfaces of constant W and lines of momentum **p**

Another way to picture the situation is to draw momentum lines (Fig. 8.06), analogous to electric field lines, which are everywhere parallel to **p** and perpendicular to the surfaces of constant W. We can describe one of these lines by giving the position vector **r** to a point on the line as a function of some parameter, which may be, for example, the distance s along the line. Since $d\mathbf{r}/ds$ is a unit vector tangential to the line, the momentum can be written $\mathbf{p}(\mathbf{r}) = p(\mathbf{r})\, d\mathbf{r}/ds$. This leads to the equation satisfied by the family of momentum lines appropriate to the particular solution $W(\mathbf{r})$,

$$p(\mathbf{r})\frac{d\mathbf{r}}{ds} = \nabla W(\mathbf{r}).$$

In the situation we are considering, the particle moves in the direction of **p** and these momentum lines form a two-parameter[11] family of possible trajectories of the particle. This set of trajectories is tied to the particular solution to the Hamilton-Jacobi equation with which we started. We can eliminate W and hence obtain the differential equation for all possible trajectories by differentiating the above equation with respect to s,

$$\frac{d}{ds}\left(p(\mathbf{r})\frac{d\mathbf{r}}{ds}\right) = \nabla\left(\frac{dW(\mathbf{r})}{ds}\right) = \nabla p(\mathbf{r}),$$

where in the second equality we have set the rate of change of W along the trajectory equal the magnitude $p(\mathbf{r})$ of the momentum.

Interestingly, this general equation for a trajectory can be obtained from a variational principle, sometimes called **Jacobi's principle**, which says:

The trajectory between points $\mathbf{r}_0$ and $\mathbf{r}_1$ at energy E
is such that the integral $\int_{\mathbf{r}_0}^{\mathbf{r}_1} p(\mathbf{r})\,ds$
is stationary.

If we parametrize the trajectory using a parameter $\lambda$, so that $\mathbf{r} = \mathbf{r}(\lambda)$ and $ds = \sqrt{\frac{d\mathbf{r}}{d\lambda}\cdot\frac{d\mathbf{r}}{d\lambda}}\,d\lambda$, this is a variational principle with independent variable $\lambda$, dependent variables **r**, and "Lagrangian" $p(\mathbf{r})\sqrt{\frac{d\mathbf{r}}{d\lambda}\cdot\frac{d\mathbf{r}}{d\lambda}}$, for which the Euler-Lagrange equation is

$$\frac{d}{d\lambda}\left(\frac{p(\mathbf{r})}{\sqrt{\phantom{x}}}\frac{d\mathbf{r}}{d\lambda}\right) = \sqrt{\phantom{x}}\,(\nabla p(\mathbf{r})).$$

If we take $\lambda = s$, then $\sqrt{\phantom{x}} = 1$ and this reduces to the previous general equation for a trajectory.

It should be emphasized that these momentum lines are possible trajectories of the particle in *space*. Nothing has yet been said about how the particle moves along the trajectories in *time*. Further, the surfaces of constant W and the momentum lines apply to some fixed energy E. If the energy is changed, the surfaces and lines change. We can introduce the time, and hence consider motion, in a couple of equivalent ways. The simpler is to observe that for the situation considered here, momentum is mass times velocity, $p = m\,ds/dt$, so

$$dt = (m/p)\,ds.$$

Another way to introduce time is to recall from our earlier discussion of the time-independent Hamilton-Jacobi equation that

---

[11]Three first order equations minus one constraint; all possible paths for a given system form a five-parameter family.

$$t = \partial W/\partial E$$

up to an additive constant. This is equivalent to the above, since

$$dt = \frac{\partial}{\partial E}\left(\frac{dW}{ds}\right) ds = \frac{\partial p}{\partial E} ds = \frac{m}{p} ds.$$

If we use the time t for the parameter $\lambda$ instead of the distance s along the trajectory, we have $\sqrt{\ } = ds/dt = p/m$ and the general equation for the trajectories becomes

$$\frac{d}{dt}\left(m\frac{d\mathbf{r}}{dt}\right) = \frac{p}{m}\nabla p = \nabla\left(\frac{p^2}{2m}\right) = -\nabla V.$$

This, of course, is Newton's second law of motion.

We can extend the idea of time dependence to the surfaces as well. The surfaces of constant W can instead be regarded as surfaces of constant

$$S(\mathbf{r},t) = W(\mathbf{r}) - Et$$

at $t = 0$. As time goes on, these surfaces of constant S move. The overall pattern of surfaces remains the same, but the surface $S = S_0$, which coincides with the surface $W = S_0$ at $t = 0$, moves so as to coincide with the surface $W = S_0 + Et$ at time t. The velocity of a point on the surface, perpendicular to the surface, can be obtained by setting

$$0 = dS/dt = \nabla W \cdot (d\mathbf{r}/dt) - E = \mathbf{p} \cdot \mathbf{v}_{surface} - E.$$

This gives $v_{surface} = E/p$, which should be contrasted with the particle velocity $v_{particle} = \partial E/\partial p = p/m$.[12] We see that these moving surfaces do not keep pace with the particles. Indeed, for fixed energy E the speeds of the surfaces and of the particles vary reciprocally. Where the surfaces of constant S are close together, they move slowly; at such points, however, the gradient of S, the momentum **p**, is large and the particles move quickly.

The above picture of a family of surfaces of constant W or S pierced by a family of perpendicular lines, which are the particle trajectories, may remind one of the situation in geometric optics, in which one has a family of surfaces of constant phase pierced by an orthogonal family of rays. As we now show, the analogy is indeed very close. Suppose we have a wave of angular frequency $\omega$ propagating through a medium which has an **index of refraction** $n(\mathbf{r}) = c/v(\mathbf{r})$ which depends on position **r**; here c is the wave speed at some reference point and $v(\mathbf{r})$ is the wave speed at point **r**. The **wave number** $k(\mathbf{r})$ is given by

$$k(\mathbf{r}) = \omega/v(\mathbf{r}) = n(\mathbf{r})\omega/c.$$

[12]The surface velocity is like phase velocity, whereas the particle velocity is like group velocity.

The wave function $\psi(\mathbf{r})$ then satisfies the time-independent **wave equation**

$$\nabla^2\psi + k^2\psi = 0.$$

Let us try a solution of the form

$$\psi = Ae^{i\phi}$$

where A and $\phi$ are the position-dependent amplitude and phase. We compute

$$\nabla\psi = (\nabla A + iA\nabla\phi)e^{i\phi} \quad \text{and} \quad \nabla^2\psi = (\nabla^2 A + 2i\nabla A\cdot\nabla\phi + iA\nabla^2\phi - A|\nabla\phi|^2)e^{i\phi}$$

and on substituting into the wave equation find

$$\nabla^2 A + 2i\nabla A\cdot\nabla\phi + iA\nabla^2\phi - A|\nabla\phi|^2 + k^2 A = 0.$$

Equating the real and imaginary parts of this separately to zero then gives the equations satisfied by the amplitude and phase,

$$\nabla^2 A - A|\nabla\phi|^2 + k^2 A = 0 \quad \text{which becomes} \quad |\nabla\phi|^2 = k^2 + \nabla^2 A/A$$

$$2\nabla A\cdot\nabla\phi + A\nabla^2\phi = 0 \quad \text{which becomes} \quad \nabla\cdot(A^2\nabla\phi) = 0.$$

These two non-linear coupled equations look considerably more complicated than the original wave equation, so one might wonder what has been achieved. The answer is that we are interested here in the geometric optics limit in which the properties of the medium change slowly with position. In particular, we assume that the distance $\ell$ over which A changes appreciably is much greater than a wavelength. Then

$$\frac{\nabla^2 A}{A} \approx \frac{1}{\ell^2} << \frac{1}{\lambda^2} = \frac{k^2}{4\pi^2},$$

and we can simplify the first equation, dropping the $\nabla^2 A/A$ on the right-hand side since it is small in comparison to $k^2$. We then obtain an equation for the phase alone,

$$|\nabla\phi(\mathbf{r})|^2 \approx k(\mathbf{r})^2.$$

This is the famous **eikonal equation** of geometric optics. We see that it has the same form as the time-independent Hamilton-Jacobi equation for a particle of mass m and energy E moving in a potential $V(\mathbf{r})$,

$$|\nabla W(\mathbf{r})|^2 = 2m(E - V(\mathbf{r})).$$

The preceding mechanics considerations can thus be translated into optics equivalents.[13] Of course, phase $\phi(\mathbf{r})$ is dimensionless, whereas $W(\mathbf{r})$ has the dimensions of "action." However, if we introduce a "constant of proportionality" $\hbar$ with the dimensions of "action," we can establish a correspondence

$$\phi(\mathbf{r}) \leftrightarrow \frac{W(\mathbf{r})}{\hbar} \qquad k(\mathbf{r}) \leftrightarrow \frac{p(\mathbf{r})}{\hbar} = \sqrt{\frac{2m}{\hbar^2}(E - V(\mathbf{r}))}\,.$$

We might now speculate, as did de Broglie and Schrödinger, that perhaps classical mechanics is in some sense the geometric optics limit of a more fundamental wave theory. With this in mind we write down the time-independent wave equation which has the Hamilton-Jacobi equation as its eikonal equation, namely

$$\nabla^2\psi + \frac{2m}{\hbar^2}(E - V(\mathbf{r}))\psi = 0\,.$$

If we take $\hbar$ to be Planck's constant (divided by $2\pi$), this, of course, is the time-independent **Schrödinger equation** of quantum mechanics.

## Exercises

**See also exercise** 11 in Chapter IV.

1. (a) Obtain Hamilton's principal function $S_H(z,t;z_0,t_0)$ for a particle of mass m which moves vertically in the uniform gravitational field g near the surface of the earth, by integrating the Lagrangian $L = \frac{1}{2}m\dot{z}^2 - mgz$ along the actual path $z = A + Bt - \frac{1}{2}gt^2$ which joins the end points. The constants A and B must be chosen so that the path passes through the end points.

   (Ans. $S_H = \frac{m}{2}\frac{(z-z_0)^2}{(t-t_0)} - mg\frac{(z+z_0)}{2} - \frac{1}{24}mg^2(t-t_0)^3$ )

   (b) Show that $S_H(z,t;z_0,t_0)$ is the type 1 generating function of a canonical transformation from present variables $(z,p_z)$ to initial variables $(z_0,p_{z0})$.

2. (a) Obtain a Jacobi complete integral $S_J(z,t;E) = W_J(z;E) - Et$ for a particle of mass m which moves vertically in the uniform gravitational field g near the surface of the earth, by integrating the time-independent Hamilton-Jacobi equation

   $$\frac{1}{2m}\left(\frac{dW}{dz}\right)^2 + mgz = E.$$

   (b) Use your solution to obtain the general solution $(z(t),p_z(t))$ to the dynamical problem.

[13]James Evans and Mark Rosenquist, ""F=ma" optics," Am. J. Phys. **54**, 876-883 (1986).

(c) Obtain Hamilton's principal function from your Jacobi complete integral by setting $S_H(z,t;z_0,t_0) = S_J(z,t;E) - S_J(z_0,t_0;E)$ and then eliminating E by using $\partial S_H/\partial E = 0$.

3. (a) Obtain Hamilton's principal function $S_H(x,t;x_0,t_0)$ for the simple harmonic oscillator by integrating the Lagrangian $L = \frac{1}{2}m\dot{x}^2 - \frac{1}{2}m\omega^2x^2$ along the actual path $x = A\sin\omega t + B\cos\omega t$ between the end points. The constants A and B must be chosen so that the path passes through the end points.
(Ans. $S_H = \dfrac{m\omega}{2\sin\omega(t-t_0)}\left[\left(x^2 + x_0^2\right)\cos\omega(t-t_0) - 2xx_0\right]$)
(b) Evaluate the action along the constant velocity path from $(x_0,t_0)$ to $(x,t)$, and compare with the result of (a). In particular, show for those paths which start at $(x_0 = 0, t_0 = 0)$ that S(actual path) < S(constant v path) provided $\omega t < \pi$.
(c) Show that $S_H(x,t;x_0,t_0)$ is the type 1 generating function of a canonical transformation from present variables $(x,t)$ to initial variables $(x_0,t_0)$.

4. The motion of a projectile near the surface of the earth (neglecting air friction) can be described by the Hamiltonian

$$H = \frac{1}{2m}\left(p_x^2 + p_y^2 + p_z^2\right) + mgz$$

where x and y denote the horizontal coordinates and z the vertical, and $p_x$, $p_y$, and $p_z$ are their conjugate momenta.
(a) Set up and find a complete integral W to the time-independent Hamilton-Jacobi equation.
(b) Use your solution to obtain x, y and z as functions of t.

5. The motion of a free particle on a plane can be described by the Hamiltonian

$$H = \frac{1}{2m}\left(p_r^2 + \frac{p_\phi^2}{r^2}\right)$$

where $p_r$ and $p_\phi$ are the momenta conjugate to the plane polar coordinates r and $\phi$.
(a) Set up and find a complete integral W to the time-independent Hamilton-Jacobi equation.
(b) Use your solution to obtain r and $\phi$ as functions of t.

6. Use the Hamilton-Jacobi method to find the general equations of motion for a three-dimensional isotropic harmonic oscillator with potential

$$V = \tfrac{1}{2}k(x^2 + y^2 + z^2) = \tfrac{1}{2}kr^2.$$

(a) First use cartesian coordinates $(x,y,z)$.
(b) Then do the problem again using spherical polar coordinates $(r,\theta,\phi)$.

7. Use the Hamilton-Jacobi method to study the motion of a particle in a dipole field with (non-central) potential

$$V = \frac{k\cos\theta}{r^2}.$$

(a) Write down the time-independent Hamilton-Jacobi equation for W in spherical polar coordinates.
(b) Show that this equation can be solved by the method of separation of variables, and obtain an expression for W of the form $W = W(r,\theta,\phi;E,\alpha_2,\alpha_3)$. Your answer will also involve certain integrals; you need not evaluate these at this stage.
(c) Interpret physically your separation constants $\alpha_2,\alpha_3$ by obtaining $p_r, p_\theta, p_\phi$ in terms of $r,\theta,\phi,E,\alpha_2,\alpha_3$. Hence show that the z-component $L_z$ of the angular momentum of the particle is constant, and further that $L^2 + 2mk\cos\theta$ is constant, where $L^2$ is the square of the total angular momentum of the particle.
(d) By considering the equation

$$\frac{\partial W}{\partial E} = t + \beta_1$$

find how r varies with time.

8. A particle of mass m moves in a field which is a superposition of a Coulomb field with potential $-k/r$ and a constant field F in the z-direction with potential $-Fz$. The total potential is

$$V = -\frac{k}{r} - Fz.$$

(a) Set up the time-independent Hamilton-Jacobi equation in paraboloidal coordinates (see exercises 3.09 and 6.04). Show that the variables separate, and obtain an expression for the Jacobi function W of the form

$$W = \sqrt{2m}\int\sqrt{k - \alpha/m - L_z^2/2m\xi^2 + F\xi^4/2 + E\xi^2}\,d\xi$$
$$+\sqrt{2m}\int\sqrt{k + \alpha/m - L_z^2/2m\eta^2 - F\eta^4/2 + E\eta^2}\,d\eta + L_z\phi$$

where $L_z$ and $\alpha$ are separation constants.
(b) Interpret physically the separation constants, showing that $L_z$ is the z-component of the angular momentum, and that $\alpha = K_z + \frac{1}{2}mF(r^2 - z^2)$ where $K_z$ is the z-component of the Laplace-Runge-Lenz vector (see exercise 1.12).

9. (a) Write down the Hamilton-Jacobi equation for a particle of mass m and charge e in an electromagnetic field described by a scalar potential $\phi$ and a vector potential $\mathbf{A}$.
(b) Show that the Hamilton-Jacobi equation is invariant under a gauge transformation,

$$\phi' = \phi - (1/c)\partial\lambda/\partial t \qquad \mathbf{A}' = \mathbf{A} + \nabla\lambda,$$

provided the Hamilton-Jacobi function is also transformed,

$$S' = S + (e/c)\lambda.$$

10. (a) Use elementary mechanics to show that the trajectory of a particle of mass m and charge e which moves in a plane $(x, y)$ perpendicular to a uniform magnetic field B is a circle, along which the particle moves with constant angular velocity $\omega = eB/mc$.
(b) Obtain Hamilton's principal function $S_H(x_1, y_1, t_1; x_0, y_0, t_0)$ by integrating the appropriate Lagrangian (in the symmetric gauge)

$$L = \tfrac{1}{2}m(\dot{x}^2 + \dot{y}^2) + \tfrac{1}{2}m\omega(x\dot{y} - y\dot{x})$$

along the path joining the end points.
(Ans. $S_H = \frac{1}{4}m\omega r^2 \cot\frac{1}{2}\omega(t_1 - t_0) + \frac{1}{2}m\omega(x_0 y_1 - x_1 y_0)$ where r is the distance between the end points)

11. (a) Write down the time-independent Hamilton-Jacobi equation for a particle of mass m and charge e in a uniform magnetic field B in the z-direction. Use cartesian coordinates and a gauge in which the vector potential is $\mathbf{A} = (-By, 0, 0)$.
(b) Show that the variables separate, and obtain a complete integral W.
(c) Use your expression for W to obtain general expressions for the cartesian coordinates as functions of time. Identify physically the separation constants $\alpha$ and their conjugate coordinates $\beta$.

12. (a) Write down the time-independent Hamilton-Jacobi equation for a particle of mass m and charge e in a uniform magnetic field B in the z-direction. Use cylindrical coordinates $(\rho, \phi, z)$ (the cartesian coordinates are $x = \rho\cos\phi$, $y = \rho\sin\phi$, z) and a gauge in which the vector potential is $\mathbf{A} = \frac{1}{2}B\rho\hat{\boldsymbol{\phi}}$ (its cartesian components are $\mathbf{A} = (-\frac{1}{2}By, +\frac{1}{2}Bx, 0)$).
(b) Show that the variables separate, and obtain a complete integral W.
(c) Use your expression for W to obtain general expressions for the cylindrical coordinates as functions of time. Identify physically the separation constants $\alpha$ and their conjugate coordinates $\beta$.

13. A particle of mass m and charge e moves in uniform crossed electric and magnetic fields, $\mathcal{E}$ in the x-direction and $\mathcal{B}$ in the z-direction.
(a) Write down the time-independent Hamilton-Jacobi equation in cartesian coordinates, and show that the variables separate for a suitable choice of gauge for the electromagnetic potentials.
(b) Use your solution to find general expressions for the cartesian coordinates of the particle as functions of time.

# CHAPTER IX

# ACTION-ANGLE VARIABLES

We now consider systems for which the Hamilton-Jacobi equation is completely separable in at least one system of coordinates. We further assume that the system is bound, so that the motion is confined to a finite region of space. Such systems include a number of physically important situations and often serve as the starting point for more complex investigations. For such systems we can introduce an especially convenient set of canonical variables called action-angle variables. We shall see that apart from their role as the natural variables for a separable system, the action-angle variables act in an interesting way under adiabatic changes to the parameters of the system.[1]

## Action-angle variables

Let $(q_1,\cdots,q_f)$ be a set of generalized coordinates in which the Hamilton-Jacobi equation is completely separable. Jacobi's complete integral W can then be written

$$W(q;\alpha) = \sum_{a=1}^{f} W_a(q_a;\alpha_1,\cdots,\alpha_f),$$

a sum of f terms each a function of a different one of the coordinates and containing f independent separation constants $(\alpha_1,\cdots,\alpha_f)$. W is the type 2 generating function of a canonical transformation to new variables in which the new momenta are these constants. For each degree of freedom the old momentum $p_a$ conjugate to the old coordinate $q_a$ is given by the first half of the transformation equations,

$$p_a = \frac{\partial W}{\partial q_a} = \frac{\partial W_a}{\partial q_a}(q_a;\alpha_1,\cdots,\alpha_f) \qquad a = 1,\cdots,f.$$

Thus, for separable systems each momentum $p_a$ can be expressed as a function of the coordinate $q_a$ conjugate to it alone, and we can consider the behavior in each degree of freedom separately. This is done most conveniently by sketching the trajectories of $p_a$ versus $q_a$ in each $(q_a,p_a)$ phase plane[2] (see Fig. 9.01).

---

[1]For parallel reading see: Max Born, *The Mechanics of the Atom*, (G. Bell and Sons, Ltd., London, 1927), translated by J. W. Fisher, revised by D. R. Hartree, pp. 45-99, 130-147; Herbert Goldstein, *Classical Mechanics*, (Addison-Wesley Publishing Company, Reading, 1980) 2nd ed., pp.457-484, 531-540.
[2]The projection of the system trajectories onto each $(q_a,p_a)$ phase plane.

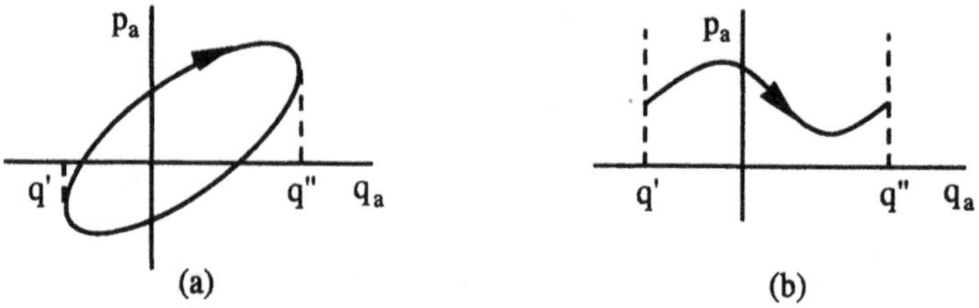


Fig. 9.01. Phase space trajectories for bound systems (a) oscillation, (b) rotation

For bound systems two types of trajectory can occur:

(a) **Oscillation**. Typically the momentum is related to its coordinate by a quadratic equation of the form $(p - a(q))^2 = b(q)$ with solutions $p = a(q) \pm \sqrt{b(q)}$. Now if $b(q)$ vanishes at $q'$ and $q''$ and is positive between these values, the resulting trajectory of p versus q is a closed loop, with q increasing from $q'$ to $q''$ for the + branch of p and decreasing from $q''$ back to $q'$ for the − branch of p. The coordinate q thus oscillates back and forth between turning points $q'$ and $q''$. In order that the + and − branches of p join smoothly at $q'$ and $q''$, we further require that $dp/dq \to \infty$ at these points, and this in turn requires $db/dq \neq 0$ at $q'$ and $q''$. In other words, the zeros of $b(q)$ should be simple.

(b) **Rotation**. Another way for the motion to remain bounded, without requiring q to lie in a finite interval, is for the system to return to its original state as the coordinate q, typically a rotation angle in this case, increases by some given amount $(q'' - q')$. Coordinates $q'$ and $q''$ are thus equivalent, so a better way to represent this situation is to replace the phase plane by a cylinder, with the lines $q = q'$ and $q = q''$ joined. The trajectories of p versus q for rotation are then also closed.

The phase plane for the simple pendulum, discussed in more detail in Chapter VI, contains both types of trajectory (Fig. 9.02). The separatrices divide the plane into three separate regions, in which occur rotation with $\dot{\theta} > 0$, oscillation, and rotation with $\dot{\theta} < 0$.

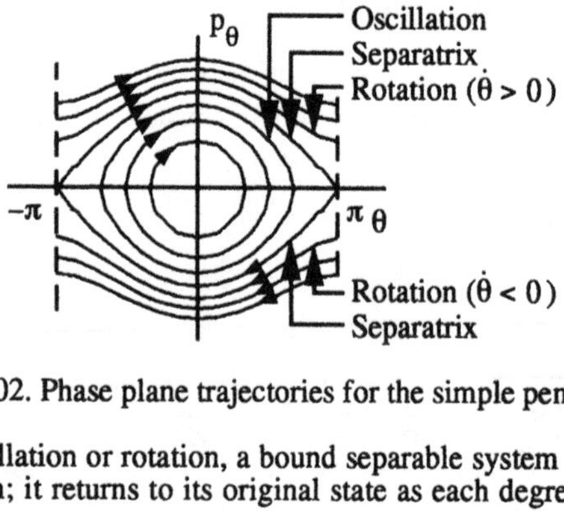


Fig. 9.02. Phase plane trajectories for the simple pendulum

For either oscillation or rotation, a bound separable system is **periodic** in each of its degrees of freedom; it returns to its original state as each degree of freedom is taken around its cycle. This, however, does not necessarily mean that the *motion* of the system is periodic. Motion has to do with the behavior in *time*. As we shall see, the various degrees of freedom do not in general move around their cycles in the same or commensurate times, so in general there is no time at which the system returns to its initial state.

Integration of the momentum $p_a$ as a function of its conjugate coordinate $q_a$ gives

$$W_a = \int p_a(q_a;\alpha_1,\cdots\alpha_f)dq_a \qquad a = 1,\cdots,f,$$

and the sum of these integrals over the various degrees of freedom gives Jacobi's complete integral W. As the coordinate $q_a$ is taken around a cycle with the other coordinates remaining fixed, the system returns to its original state, and $W_a$ changes by

$$\Delta W_a = \oint p_a\, dq_a$$

where $\oint$ denotes integration around the cycle. This means that $W_a$, and hence Jacobi's complete integral W, is a multivalued function of the state of the system. The integral is the area in the $(q_a, p_a)$ phase plane enclosed by (or for rotation, under one cycle of) the trajectory. We denote the change in W by $2\pi I_a$, where

$$I_a = \frac{1}{2\pi}\oint p_a\, dq_a$$

is called the **action variable** for the a'th degree of freedom. Action variables have the dimensions of "action." The above definition gives them in terms of the separation constants $(\alpha_1,\cdots,\alpha_f)$,

$$I_a = I_a(\alpha_1,\cdots,\alpha_f) \qquad a = 1,\cdots,f.$$

We assume that this set of f equations can be inverted to obtain the $\alpha$'s in terms of the I's. We can then take the more fundamental action variables $(I_1,\cdots,I_f)$ as the new momenta, rather than the constants $(\alpha_1,\cdots,\alpha_f)$ which the process of separation of variables happens to give. Jacobi's complete integral becomes

$$W(q;I) = \sum_{a=1}^{f} W_a(q_a;I_1,\cdots,I_f),$$

where we have replaced the $\alpha$'s by their expressions in terms of the I's.

For each degree of freedom the coordinate $\phi_a$ conjugate to the action variable $I_a$ is now given by the second half of the transformation equations,

$$\phi_a = \frac{\partial W}{\partial I_a}(q_1,\cdots,q_f;I_1,\cdots,I_f) \qquad a = 1,\cdots,f.$$

These coordinates are called the **angle variables**. The function $W(q;I)$ is thus the generating function of a canonical transformation from variables $(q,p)$ to **action-angle variables** $(\phi, I)$. The angle variables are dimensionless. Further, if the a'th degree of freedom is taken around a cycle with the other degrees remaining fixed, W changes by $2\pi I_a$, and the angle variable $\phi_a$ for this degree of freedom changes by $2\pi$ while the other angle variables remain unchanged,

$$\Delta_a \phi_b = 2\pi\delta_{ab}.$$

The state of the system is a periodic function of each of the angle variables with period $2\pi$. This means that any single-valued function of the state of the system, such as a (non-rotation) coordinate x, can be expressed as a multiple Fourier series in the angle variables, of the form

$$x = \sum_{n_1=-\infty}^{\infty} \cdots \sum_{n_f=-\infty}^{\infty} A_{n_1,\cdots,n_f}(I_1,\cdots,I_f)e^{i(n_1\phi_1+\cdots+n_f\phi_f)}$$

where the n's are (positive and negative) integers.

Let us now look at the time dependence of the variables. The Hamiltonian, which separation of variables gives as a function of the separation constants $(\alpha_1,\cdots,\alpha_f)$, can now be written as a function of the action variables $(I_1,\cdots,I_f)$ alone, $H = H(I_1,\cdots,I_f)$. The angle variables are thus all cyclic variables. The action variables are constant in time, and the angle variables change with time according to

$$\frac{d\phi_a}{dt} = \frac{\partial H(I)}{\partial I_a} = \omega_a(I) \qquad a = 1,\cdots,f.$$

These equations define a set of constant in time (but action-dependent) **angular frequencies** $\omega_a(I)$. The equations can be integrated immediately to give

$$\phi_a(t) = \phi_a(0) + \omega_a(I)t \qquad a = 1,\cdots,f.$$

The angle variables increase uniformly with time. Any single-valued function x of the state of the system can now be written

$$x(t) = \sum_{n_1=-\infty}^{\infty} \cdots \sum_{n_f=-\infty}^{\infty} \tilde{A}_{n_1,\cdots,n_f}\, e^{i(n_1\omega_1+\cdots+n_f\omega_f)t}$$

where the amplitudes $\tilde{A}$ and frequencies $\omega$ are functions of the (constant) actions I.

The motion of a system with f degrees of freedom is described in general by specifying the trajectory of the system in 2f-dimensional phase space and the way the system moves along this trajectory. Separable systems, however, possess a set of f single-valued constants of the motion, the action variables $(I_1,\cdots,I_f)$, so their motions are comparatively simple, a given motion taking place in an f-dimensional subspace of phase space, labeled by the action variables. One can show that the topology of this subspace is usually that of an f-dimensional **torus**.[3] Points on a given torus are labeled by the angle variables $(\phi_1,\cdots,\phi_f)$. The f independent cycles of the system correspond to the f topologically distinct circuits of the torus. For a system with one degree of freedom, the tori are the closed curves $H(q,p) = \text{constant}$. For a separable system with two degrees of freedom, the tori are two-dimensional surfaces, the surface of a "doughnut" (Fig. 9.03).

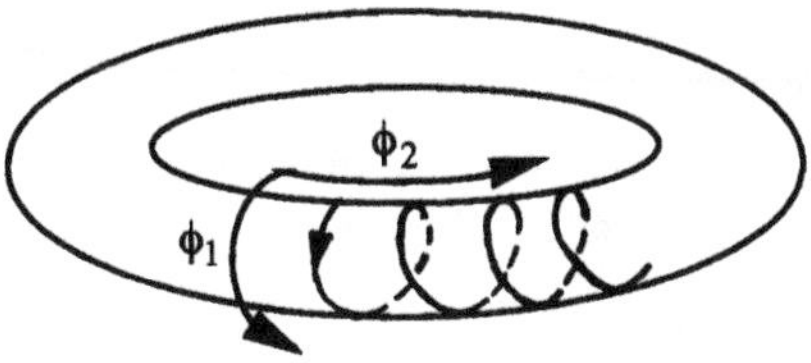


Fig. 9.03. Constant action torus in phase space

The system trajectory winds around and around the torus. The nature of the trajectory depends on the frequencies $\omega_a = \partial H/\partial I_a$. If the frequencies are all incommensurate, the trajectory never returns to its initial point and it eventually covers the torus densely. On the other hand, if the ratio of any two of the frequencies is a rational number or, more generally, if the frequencies are related by one or more linear equations of the form

$$n_1\omega_1 + n_2\omega_2 + \cdots + n_f\omega_f = 0$$

with integer coefficients $n_a$, the trajectory is restricted to a subspace of the torus. In particular, if there are $f-1$ independent such relations, the subspace is one-dimensional and the trajectory is a closed curve. In this latter case the frequency ratios $\omega_a/\omega_b$ are all rational numbers, and the frequencies can all be written as integral multiples of some frequency $\omega_0$. In a time interval $2\pi/\omega_0$ all angle variables increase by integral multiples of $2\pi$ and the system returns to its original state. The motion is then **periodic**. Since the frequencies are functions of the action variables, whether or not such relations between the frequencies exist and, in particular, whether or not the motions are periodic depends on the torus. In the general situation in which the frequencies are functionally independent ($\det|\partial\omega/\partial I| = \det|\partial^2 H/\partial I^2| \neq 0$) and for most tori there are no such relations. Nevertheless, distributed among the tori is an everywhere dense but of measure zero set of tori on which there are rational relations between the frequencies and, in particular, on which the motions are periodic.

[3]V.I. Arnol'd, *Mathematical Methods of Classical Mechanics*, (Springer-Verlag New York, Inc. 1978), trans. K. Vogtmann and A. Weinstein, section 49; P. J. Richens and M. V. Berry, "Pseudointegrable Systems in Classical and Quantum Mechanics," Physica **2D**, 495-512 (1981).

It sometimes happens that the frequencies satisfy one or more linear equations of the above form *independent* of the values of the action variables. If they satisfy s such relations we say that the system is s-fold **degenerate**. In particular, if the frequencies satisfy $f-1$ such relations, or equivalently if the ratio of every pair of the frequencies is a rational number, we say that the system is $f-1$ fold or **completely degenerate**. Degeneracy has a number of consequences. To begin with, there are additional single-valued constants of motion over and above the f action variables $I_a$ possessed by any separable system. Since $\omega_a = d\phi_a/dt$, where the $\phi_a$ are the angle variables, the above linear relation among the frequencies can be integrated to give

$$n_1\phi_1 + n_2\phi_2 + \cdots + n_f\phi_f = \text{constant}.$$

The left-hand side is a constant of the motion, but it is not single-valued. However, it only changes by integral multiples of $2\pi$ as the $\phi_a$'s are taken around their cycles, so its sine or cosine *is* a new single-valued constant of the motion. The constant of the motion $G(n_1\phi_1 + n_2\phi_2 + \cdots + n_f\phi_f)$ generates an infinitesimal canonical transformation

$$I'_a = I_a - \varepsilon G' n_a \qquad a = 1,\cdots,f$$

which leaves the Hamiltonian invariant. Thus only those combinations of the action variables which are invariant under this transformation can appear in the Hamiltonian.[4] Finally, consider the connection between degeneracy and the possibility of separating the Hamilton-Jacobi equation in more than one coordinate system. We have already noted that the trajectory in phase space for a separable non-degenerate system covers the torus defined by the action variables densely. This implies that the trajectory in real space fills the region $(q'_1 < q_1 < q''_1, \cdots, q'_f < q_f < q''_f)$ defined by the (action-dependent) turning points of the coordinates. It is then clear that in this case separation of variables is possible in only one system of coordinates, that defined by the region filled by the trajectory. The trajectory for a degenerate system does not fill a region and thereby define a coordinate system, and separation of variables may be possible in more than one coordinate system.

## Example: simple harmonic oscillator

We have seen that Jacobi's complete integral for a simple harmonic oscillator of mass m and frequency $\omega$ is

$$W = (E/\omega)(\phi + \sin\phi\cos\phi),$$

where E is the energy, and the phase angle $\phi$ is related to the coordinate x and momentum p by

$$x = \sqrt{2E/m\omega^2}\,\sin\phi \qquad p = \sqrt{2mE}\,\cos\phi.$$

[4]In two degrees of freedom the appropriate combination is $n_2 I_1 - n_1 I_2$.

As the system is taken around a complete cycle, the phase angle $\phi$ increases by $2\pi$ and W increases by

$$\Delta W = 2\pi E/\omega .$$

The action variable I is thus given by

$$I = E/\omega .$$

The trajectory in phase space, which for this one-degree-of-freedom situation is the "torus," is an ellipse with semi-x-axis $\sqrt{2E/m\omega^2}$, semi-p-axis $\sqrt{2mE}$, and hence an enclosed area $\pi \times \sqrt{2E/m\omega^2} \times \sqrt{2mE} = 2\pi E/\omega$ (Fig. 9.04). This, divided by $2\pi$, is another way to obtain the action variable I.

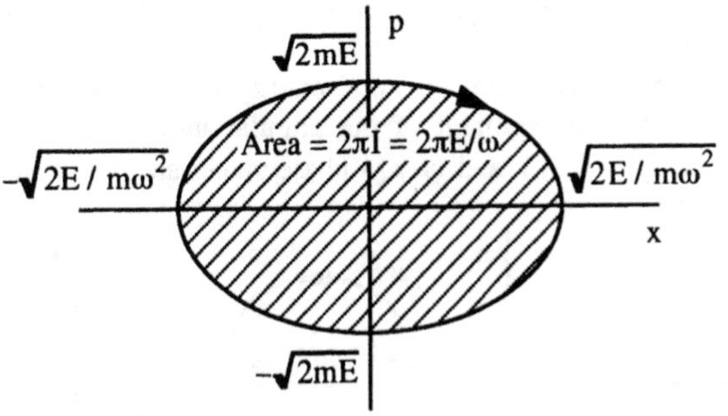


Fig. 9.04. Phase space trajectory for a simple harmonic oscillator

We now, in the expression for W, replace the energy E by its expression in terms of the action variable I, obtaining

$$W(x,I) = I(\phi + \sin\phi\cos\phi) \quad \text{where} \quad \sin\phi = x\sqrt{m\omega/2I}.$$

W(x,I) is the generating function of a canonical transformation to new variables in which the new momentum is the action variable I. The coordinate conjugate to I, the angle variable, is given by

$$\left(\frac{\partial W}{\partial I}\right)_x = \phi + \sin\phi\cos\phi + I(2\cos^2\phi)\left(\frac{\partial\phi}{\partial I}\right)_x = \phi$$

where we have made use of $(\partial\phi/\partial I)_x = -\tan\phi/2I$. The angle variable is thus simply the phase angle $\phi$. Its time dependence is given by Hamilton's equation with Hamiltonian $H = \omega I$,

$$\frac{d\phi}{dt} = \frac{\partial H}{\partial I} = \omega.$$

So $\phi$ increases uniformly at a rate $\omega$, which for the simple harmonic oscillator is action independent.

## Example: central force

As we have seen, the Hamilton-Jacobi equation for a particle of mass m moving in a spherically symmetric potential V(r) is completely separable in spherical polar coordinates $(r,\theta,\phi)$. The momenta conjugate to these coordinates are given by

$$p_r = \sqrt{2m}\sqrt{E - V(r) - \frac{L^2}{2mr^2}}, \qquad p_\theta = \sqrt{L^2 - \frac{L_z^2}{\sin^2\theta}}, \qquad p_\phi = L_z$$

where the separation constants are the total energy E, the magnitude of the angular momentum L, and the z-component of the angular momentum $L_z$. The resulting trajectories in the $(r, p_r)$, $(\theta, p_\theta)$, and $(\phi, p_\phi)$ phase planes are shown in Fig. 9.05.

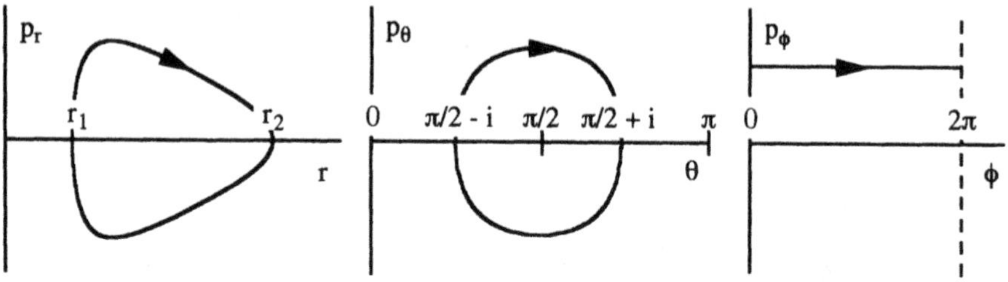


Fig. 9.05. Phase space trajectory for a central force

The r and $\theta$ degrees of freedom are oscillatory, with r oscillating between $r_1$ and $r_2$ (where $r_1$ and $r_2$ are appropriate roots of $V_{eff}(r) = V(r) + L^2/2mr^2 = E$), and with $\theta$ oscillating between $\pi/2 - i$ and $\pi/2 + i$ (where $\cos i = L_z/L$),[5] whereas the $\phi$ degree is rotational, with $\phi$ changing by $2\pi$ over a cycle. The action variables are given by

$$I_r = \frac{\sqrt{2m}}{2\pi}\oint\sqrt{E - V(r) - \frac{L^2}{2mr^2}}\,dr, \quad I_\theta = \frac{1}{2\pi}\oint\sqrt{L^2 - \frac{L_z^2}{\sin^2\theta}}\,d\theta, \quad I_\phi = \frac{1}{2\pi}\oint L_z\,d\phi,$$

[5]Here we have assumed that $L_z \geq 0$ and hence $0 \leq i \leq \pi/2$; if $L_z < 0$ and $\pi/2 < i \leq \pi$, then the limits of $\theta$ are $\pi/2 - (\pi - i)$ and $\pi/2 + (\pi - i)$.

the integrations being over one cycle.

The integration for $I_\phi$ is trivial, since $L_z$ is constant. For $L_z$ positive $\phi$ increases by $2\pi$ over a cycle, whereas for $L_z$ negative $\phi$ decreases by $2\pi$. Thus

$$I_\phi = |L_z|.$$

The modulus sign here arises because we have chosen to *define* the action variable by going around the cycle in the direction the system would develop in time.

The integration for $I_\theta$ can be performed by writing the integrand $\sqrt{\ } = (\sqrt{\ })^2/\sqrt{\ }$, thus obtaining

$$I_\theta = \frac{L}{2\pi}\oint \frac{d\theta}{\sqrt{1 - \dfrac{\cos^2 i}{\sin^2\theta}}} - \frac{L_z}{2\pi}\oint \frac{\cos i\, d\theta}{\sin^2\theta\sqrt{1 - \dfrac{\cos^2 i}{\sin^2\theta}}}.$$

These integrals, in indefinite form, have both been evaluated previously in Chapter VIII. Using those results, we find

$$I_\theta = \frac{L}{2\pi}\oint d\bar{\theta} - \frac{L_z}{2\pi}\oint d\bar{\phi}$$

where $\sin\bar{\theta} = \cos\theta/\sin i$ and $\sin\bar{\phi} = \cot i \cot\theta$. Over a cycle $\bar{\theta}$ increases by $2\pi$, whereas $\bar{\phi}$ increases by $2\pi$ if $L_z$ is positive or decreases by $2\pi$ if $L_z$ is negative. We thus obtain

$$I_\theta = L - |L_z|.$$

The modulus sign on $L_z$ is essential (as a check, note that the defining equation shows that $I_\theta$ is even in $L_z$). The overall sign of $I_\theta$, however, depends on how we choose to go around the cycle. Here we have gone around in the direction the system would develop in time. The action variable $I_\theta$ is then positive (note $|L_z| \le L$) or zero if the orbit lies in the x-y plane ($|L_z| = L$). In this latter case the trajectory in the $(\theta, p_\theta)$ phase plane degenerates to the point $(\theta = \pi/2, p_\theta = 0)$.

The integration for $I_r$ cannot be performed explicitly without specifying the potential $V(r)$, but we *can* see that $I_r$ depends on the separation constants E and L, but not on $L_z$. To put it another way, the energy (or Hamiltonian) can be expressed as a function of the action variable $I_r$ and the combination of action variables $L = I_\theta + I_\phi$; the variables $I_\theta$ and $I_\phi$ do not appear separately. This means that the frequencies $\omega_\theta = \partial E/\partial I_\theta$ and $\omega_\phi = \partial E/\partial I_\phi$ are equal, and the system is at least one-fold degenerate. This is also clear from the fact that the orbits lie in a plane through the origin: as $\phi$ goes through one cycle, $\theta$ also goes through one cycle. As a particular case, now consider the gravitational (or attractive electrostatic) force with potential $V(r) = -k/r$. To perform the integration for $I_r$, again rewrite the integrand using $\sqrt{\ } = \left(\sqrt{\ }\right)^2/\sqrt{\ }$ to obtain

$$\frac{2\pi I_r}{\sqrt{2m}} = \oint \frac{(E + k/2r)}{\sqrt{E + k/r - L^2/2mr^2}}dr + \oint \frac{k/2r}{\sqrt{\ }}dr - \oint \frac{L^2/2mr^2}{\sqrt{\ }}dr.$$

The first term is zero, since the integrand is the exact differential $d\sqrt{Er^2 + kr - L^2/2m}$. The second term simplifies with the transformation $r/a = 1 - e\cos\psi$ where $\psi$ is the eccentric anomaly. The third simplifies with the transformation $a(1-e^2)/r = 1 + e\cos\alpha$ where $\alpha$ is the true anomaly. Here $a = k/(-2E)$ is the semi-major axis and $e = \sqrt{1 + 2L^2E/mk^2}$ is the eccentricity of the orbit (see Chapter I). We find

$$2\pi I_r = \sqrt{mk^2/(-2E)}\oint d\psi - L\oint d\alpha.$$

Both $\psi$ and $\alpha$ increase by $2\pi$ over a cycle, so we obtain

$$I_r = \sqrt{mk^2/(-2E)} - L.$$

This can be turned around to express the energy in terms of the action variables, thus

$$E = -\frac{mk^2}{2(I_r + I_\theta + I_\phi)^2}.$$

So for this potential the energy depends only on the combination of action variables $I_r + I_\theta + I_\phi$. This means that the frequencies $\omega_r = \partial E/\partial I_r$, $\omega_\theta = \partial E/\partial I_\theta$, and $\omega_\phi = \partial E/\partial I_\phi$ are all equal, and this system is completely degenerate. The orbits are closed and, further, for this particular case r, $\theta$, and $\phi$ go through their cycles at the same rate.

Let us now consider the angle variables. To find these, we take Jacobi's complete integral $W(r,\theta,\phi;E,L,L_z)$, express the separation constants in terms of the action variables, $E = E(I_r, I_\theta + I_\phi)$, $L = I_\theta + I_\phi$, $L_z = \pm I_\phi$, and evaluate the derivatives

$$\frac{\partial W}{\partial I_r} = \frac{\partial W}{\partial E}\omega_r, \qquad \frac{\partial W}{\partial I_\theta} = \frac{\partial W}{\partial E}\omega_\theta + \frac{\partial W}{\partial L}, \qquad \frac{\partial W}{\partial I_\phi} = \frac{\partial W}{\partial E}\omega_\phi + \frac{\partial W}{\partial L} \pm \frac{\partial W}{\partial L_z}.$$

These are the angle variables conjugate to the action variables $I_r$, $I_\theta$, and $I_\phi$. We have already noted that for any central force the frequencies $\omega_\phi$ and $\omega_\theta$ are equal. This implies that the difference of the angle variables $\partial W/\partial I_\phi$ and $\partial W/\partial I_\theta$ is a constant of the motion. In fact we have

$$\frac{\partial W}{\partial I_\phi} - \frac{\partial W}{\partial I_\theta} = \pm\frac{\partial W}{\partial L_z}.$$

The right-hand side was identified in Chapter VIII as $\phi - \bar{\phi} = \beta_{L_z}$, the longitude of the ascending node. Its sine or cosine gives a single-valued constant of the motion. Together with the constants $I_\theta$ and $I_\phi$, it can be used, for example, to construct the three components of the (constant) angular momentum vector. For the gravitational force the frequencies $\omega_\theta$ and $\omega_r$ are also equal, and the difference of the angle variables $\partial W/\partial I_\theta$ and $\partial W/\partial I_r$ is a second constant of the motion. We have

$$\frac{\partial W}{\partial I_\theta} - \frac{\partial W}{\partial I_r} = \frac{\partial W}{\partial L}.$$

The right-hand side was identified in Chapter VIII as $\beta_L$, the argument of perihelion. Basically, it specifies a direction in the orbital plane. Finally, turning to the angle variable $\partial W/\partial I_r$, we see from the results in Chapter VIII that it equals $\omega_r(\beta_E + t)$, the mean anomaly.

## Adiabatic change

The Hamiltonian of a mechanical system usually depends on various parameters $\mathbf{X} = (X_1, X_2, \cdots)$ in addition to the canonical variables q and p. For example, the Hamiltonian for a simple plane pendulum depends on the mass m of the bob, the length $\ell$ of the string, and the gravitational field g. We are interested here in what happens if these parameters are changed slowly with time, and in a way uncorrelated with the motion of the system. Such a change to a system is called an **adiabatic** change. The behavior of the action-angle variables under adiabatic changes turns out to be especially interesting. For simplicity, we consider only systems with one degree of freedom.

We begin with a couple of simple examples. First, let us look again at the simple plane pendulum. Suppose that the pendulum is undergoing small amplitude oscillations, $\theta = \theta_0 \sin\omega t$, of frequency $\omega = \sqrt{g/\ell}$ and amplitude $\theta_0$, and that we adiabatically shorten the length of the string by pinching it at the top and then sliding the pinch downwards (Fig. 9.06).[6] The frequency of oscillation increases. Further, we do work on the pendulum, so its energy $E = \frac{1}{2}m\omega^2(\ell\theta_0)^2 = \frac{1}{2}mg\ell\theta_0^2$ and amplitude of oscillation increase. To find how these changes are related, we now turn to the dynamics.

[6]This avoids the raising of the point of equilibrium which would accompany pulling the string up through a hole in the ceiling.

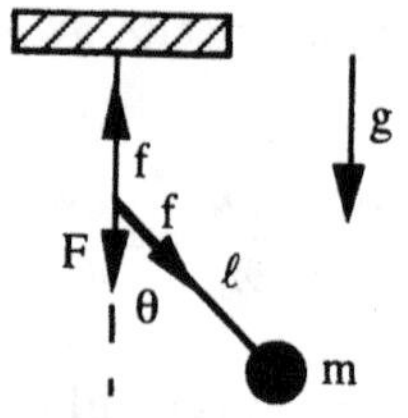


Fig. 9.06. Pendulum with varying length

The downward force we must exert on the pinch is

$$F = f(1-\cos\theta) \approx \tfrac{1}{2} f\,\theta^2$$

where $f \approx mg$ is the tension in the string. The force we exert, averaged over a period of the oscillation, is thus

$$\langle F\rangle \approx \tfrac{1}{2} mg\,\theta_0^2 \langle \sin^2 \omega t\rangle = \tfrac{1}{4} mg\,\theta_0^2 = E/2\ell .$$

The work we do in sliding the pinch down a small distance $-\Delta\ell$ in a time interval long in comparison with the period is $-\langle F\rangle \Delta\ell$ and equals the change in energy of the pendulum,

$$\Delta E = -\langle F\rangle \Delta\ell = -E\,\Delta\ell/2\ell .$$

Thus, on integrating we see that as the length of the pendulum decreases, the energy increases such that

$$E\sqrt{\ell} = \text{constant}.$$

This quantity, proportional to the action variable $I = E/\omega$, remains unchanged under an adiabatic change in the length of the pendulum. It is an **adiabatic invariant**. We shall soon show that this is a general property of action variables.

Before doing so, however, it is natural to ask: what is the change in the angle variable if the parameters of a system are changed adiabatically? Surprisingly, this question does not appear to have been considered, at least in a general way, until Hannay,[7] inspired by closely related work of Berry[8] on the change in the phase of a quantum wave function under adiabatic changes, studied the problem. We have seen that if the parameters **X** are constant, the change in the angle variable in a time interval $t_0$ to $t$ is $\omega(t-t_0)$, where $\omega = \partial H/\partial I$ is the angular frequency. This frequency depends on the parameters **X** of the system, and if these are changed adiabatically with time, we might

[7] J. H. Hannay, "Angle variable holonomy in adiabatic excursion of an integrable Hamiltonian," J. Phys. A **18**, 221-230 (1985); M. V. Berry, "Classical adiabatic angles and quantal adiabatic phase," J. Phys. A **18**, 15-27 (1985).

[8] M. V. Berry, "Quantal phase factors accompanying adiabatic changes," Proc. R. Soc. Lond. A **392**, 45-57 (1984).

expect the change in the angle variable to become $\int_{t_0}^{t} \omega(\mathbf{X}(t))\,dt$. This, however, is not always the case; there is sometimes an additional change, now called the **Hannay angle**.

A simple system which illustrates this is the **Hannay hoop**: a bead slides without friction around a plane wire hoop of arbitrary shape (Fig. 9.07). We describe the location of the bead by giving its displacement s along the hoop from some fixed point on the hoop. The magnitude p of the momentum of the bead is constant. The action-angle variables are $\phi = 2\pi s/\ell$ and $I = p\ell/2\pi$ where $\ell$ is the total distance around the hoop.

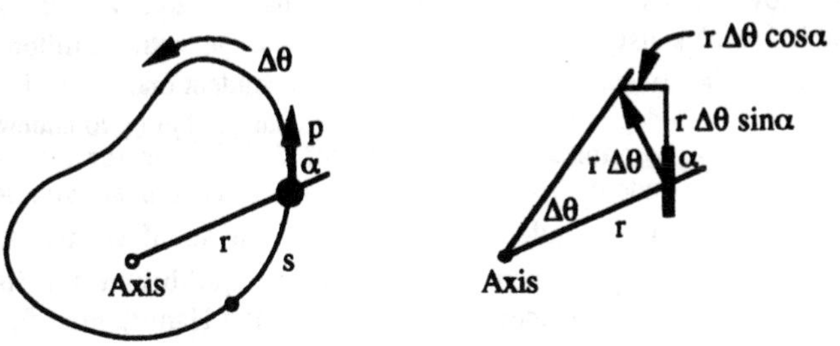


Fig. 9.07. Hannay hoop

Now let the hoop be slowly rotated in its own plane about some axis perpendicular to the plane. At a given instant suppose that the bead is at a part of the hoop which is a distance r from the axis, and which makes an angle α with r. If the hoop is rotated through a small angle $\Delta\theta$, this part of the hoop moves a distance $r\Delta\theta$ perpendicular to r. This displacement can be resolved into components $r\Delta\theta\cos\alpha$ perpendicular and $r\Delta\theta\sin\alpha$ parallel to the hoop. The hoop exerts a force on the bead perpendicular to the hoop and thus carries the bead along with it through the perpendicular displacement. However, it does *not* exert a force on the bead parallel to the hoop, so the bead slips "backwards" relative to the hoop a displacement $\Delta s = -r\Delta\theta\sin\alpha$. Now the bead is actually moving around the hoop and, in the course of the small rotation $\Delta\theta$ of the hoop, it makes many circuits. We thus average this slippage over all the various locations of the bead around the hoop,

$$\langle \Delta s \rangle = -\left(\frac{1}{\ell}\int_0^{\ell} r\sin\alpha\,ds\right)\Delta\theta.$$

The integrand is $2\,dA$ where dA is the area swept out by r as we go from s to $s + ds$. We thus find

$$\langle \Delta s \rangle = -(2A/\ell)\Delta\theta$$

where A is the total area enclosed by the hoop. If the hoop is slowly rotated through a complete rotation so as to return to its original position, $\Delta\theta = 2\pi$ and the bead slips $\langle \Delta s \rangle = -4\pi A/\ell$. The bead is behind, by this amount, where it would be if the hoop had not been rotated. This displacement, or rather the slippage $\langle\Delta\phi\rangle = -8\pi^2 A/\ell^2$ in the corresponding angle variable, is the Hannay angle. If the hoop is circular $\langle\Delta\phi\rangle = -2\pi$, which is easily understood, at least if the axis of rotation is at the center of the hoop. For other shapes of hoop we can write $\langle\Delta\phi\rangle = -2\pi + 2\pi(1 - 4\pi A/\ell^2)$, with the factor in brackets being always positive.

Let us now turn to the general one-dimensional mechanical system, described by canonical variables $(q,p)$ and with Hamiltonian $H_0(q,p,\mathbf{X})$. Initially the parameters $\mathbf{X}$ are constant. Now suppose that the parameters are changed adiabatically, and ***assume*** that the variables $(q,p)$ satisfy Hamilton's canonical equations with Hamiltonian obtained simply by replacing the constant parameters by time-dependent ones, thus $H_0(q,p,\mathbf{X}(t))$. Whether or not this is so depends on the system and changes being considered, and also on the set of canonical variables used. For example, it is true for the simple plane pendulum with time-dependent string length $\ell$ if we use as canonical variables the angle $\theta$ from the vertical and the angular momentum $p_\theta$, but not if we use as canonical variables the horizontal displacement $x = \ell\sin\theta$ from equilibrium and its conjugate momentum $p = p_\theta/\ell\cos\theta$.[9] For these latter variables the Hamiltonian for the simple plane pendulum with time-dependent string length contains the additional term $(px/\ell)(d\ell/dt)$; it depends on the rate of change of $\ell$.

We are interested in the equations of motion satisfied by the action-angle variables $(\phi,I)$. At each instant t these variables are obtained from the $(q,p)$ variables at that instant by a time-dependent canonical transformation, with generating function $W(q,I,\mathbf{X}(t))$ which depends on the parameters $\mathbf{X}(t)$ at that instant. The action-angle variables thus satisfy Hamilton's canonical equations with Hamiltonian

$$H(\phi,I,\mathbf{X}(t)) = H_0(I,\mathbf{X}(t)) + \left(\frac{\partial W(q,I,\mathbf{X}(t))}{\partial \mathbf{X}}\right)_{q,I} \cdot \frac{d\mathbf{X}(t)}{dt},$$

where $H_0(I,\mathbf{X}(t))$ is the Hamiltonian for the $(q,p)$ variables expressed in terms of the action-angle variables. Because of the nature of these latter variables, it depends only on the action variable I and not on the angle variable. In the second term we must express q in terms of the action-angle variables $(\phi,I)$ after differentiating W with respect to the parameters $\mathbf{X}$. We see that the Hamiltonian for the action-angle variables usually depends on the rate of change $d\mathbf{X}(t)/dt$ of the parameters as well as on the parameters themselves. Now we have

$$\left(\frac{\partial W}{\partial \mathbf{X}}\right)_{\phi,I} = \left(\frac{\partial W}{\partial q}\right)_{I,\mathbf{X}}\left(\frac{\partial q}{\partial \mathbf{X}}\right)_{\phi,I} + \left(\frac{\partial W}{\partial \mathbf{X}}\right)_{q,I} = p\left(\frac{\partial q}{\partial \mathbf{X}}\right)_{\phi,I} + \left(\frac{\partial W}{\partial \mathbf{X}}\right)_{q,I},$$

so the canonical equations of motion for the action-angle variables can be written

---

[9] The variables $(x,p)$ are obtained from the variables $(\theta,p_\theta)$ by a canonical transformation generated by $F_2 = p\ell\sin\theta$; note that p is *not* the x-component of the linear momentum.

$$\frac{d\phi}{dt} = \omega(I,\mathbf{X}) + \frac{\partial}{\partial I}(-p\nabla q + \nabla W)\cdot\frac{d\mathbf{X}}{dt} \qquad \frac{dI}{dt} = -\frac{\partial}{\partial\phi}(-p\nabla q + \nabla W)\cdot\frac{d\mathbf{X}}{dt}.$$

Here $\omega(I,\mathbf{X}) = \partial H_0(I,\mathbf{X})/\partial I$ is the angular frequency, and we now use $\nabla = \partial/\partial\mathbf{X} = (\partial/\partial X_1, \partial/\partial X_2, \cdots)$ to denote differentiation with respect to the parameters while keeping the action-angle variables fixed. These equations hold no matter how the parameters change with time. If they change slowly so that $(1/\mathbf{X})(d\mathbf{X}/dt) << \omega$, the system goes through many cycles in the time it takes the parameters to change appreciably; it undergoes an adiabatic change. In calculating the change in the action-angle variables over many cycles we can then to a good approximation replace the right-hand sides of the equations of motion by their average

$$\langle\cdots\rangle = \frac{1}{2\pi}\int_0^{2\pi}\cdots d\phi$$

over a cycle.

Consider first the "action" equation. Since the term in brackets on the right is a single-valued function of $\phi$,[10] the averaging gives zero and we have for the behavior of the action variable under adiabatic changes

$$\frac{dI}{dt} \approx 0.$$

The action variable is constant. It is an **adiabatic invariant**.

Now consider the "angle" equation. The averaging of the right-hand side of this equation over a cycle gives

$$\frac{d\phi}{dt} \approx \omega(I,\mathbf{X}) + \frac{\partial \mathcal{A}(I,\mathbf{X})}{\partial I}\cdot\frac{d\mathbf{X}}{dt}$$

where

$$\mathcal{A}(I,\mathbf{X}) = -\langle p\nabla q\rangle + \langle \nabla W\rangle.$$

Recalling that the action variable I is constant, integration with respect to time then gives the change in the angle variable in the interval $t_0$ to t,

$$\Delta\phi \approx \int_{t_0}^{t}\omega(I,\mathbf{X}(t))\,dt + \frac{\partial}{\partial I}\int_{\mathbf{X}_0}^{\mathbf{X}}\mathcal{A}(I,\mathbf{X})\cdot d\mathbf{X}.$$

The first term is the expected dynamical change in the angle variable. The second, which depends on the path in parameter space but not on the time, is the Hannay change. Now

---

[10]The function W is multi-valued, but $\nabla W$ is single-valued since the change $2\pi I$ in W around a cycle does not depend on $\mathbf{X}$.

the angle variable is defined only up to a canonical transformation (generated by $F_2 = \phi I' + \Lambda(I',\mathbf{X})$)

$$\phi' = \phi + \frac{\partial\Lambda(I,\mathbf{X})}{\partial I} \qquad \text{(together with } I' = I\text{)}.$$

That is, the "zero," the point from which we measure the angle variable, may be chosen arbitrarily. Further, this choice may depend on the action variable. As can be seen from the angle equation, this transformation changes the quantity $\mathcal{A}$ to

$$\mathcal{A}' = \mathcal{A} + \nabla\Lambda.$$

This has the same form as a gauge transformation of the vector potential in electrodynamics; hence the suggestive notation. The Hannay change in the angle variable is thus usually "gauge" dependent. For *closed* circuits in parameter space, however, the Hannay change

$$\Delta\phi_H = \frac{\partial}{\partial I}\oint \mathcal{A}(I,\mathbf{X})\cdot d\mathbf{X}$$

is "gauge" *independent*. We call this quantity the **Hannay angle**. The integral of the one-form $\mathcal{A}(I,\mathbf{X})\cdot d\mathbf{X}$ around a closed circuit in parameter space can be rewritten, by using Stokes' theorem, as an integral of the appropriate two-form over a surface which has the circuit as its edge.[11] If, for simplicity, we take a system with three parameters, we can use ordinary vector notation, introducing

$$\mathcal{B} = \nabla\times\mathcal{A} = \langle\nabla q\times\nabla p\rangle,$$

a quantity analogous to the magnetic field. Like its electromagnetic counterpart, it is invariant under "gauge" transformations. The Hannay angle can then be expressed in terms of the "flux" of this "magnetic field" through a surface in parameter space which spans the circuit,

$$\Delta\phi_H = \frac{\partial}{\partial I}\iint \mathcal{B}(I,\mathbf{X})\cdot d\mathbf{S};$$

this is an especially elegant way to state the result.

In our earlier treatment of the Hannay hoop we worked directly with the action-angle variables. If we wish instead to use the above formulas, we must first describe the system using canonical variables whose equations of motion contain only the time-dependent parameters $\mathbf{X}(t)$ themselves and not their time derivatives $d\mathbf{X}(t)/dt$. Suitable variables for the Hannay hoop are

---

[11] See Chapter VIII.

$$Q = \int_0^s \frac{ds}{r\sin\alpha} + \theta \quad \text{and} \quad P = pr\sin\alpha,$$

where the parameter $\theta(t)$ is the angle turned through.[12] If this is changed by a small amount $\Delta\theta$, the canonical variables (Q,P) change by

$$\Delta Q = \frac{\Delta s}{r\sin\alpha} + \Delta\theta = -\Delta\theta + \Delta\theta = 0 \quad \text{and} \quad \Delta P = 0;$$

that is, they remain constant as required. To use the above expression for the Hannay angle, we need "$\nabla Q$" $= \partial Q/\partial\theta = 1$ and "$P\nabla Q$" $= (2\pi I/\ell) r\sin\alpha$, where we have expressed p in terms of the action variable I. We then have $\mathcal{A} = -\langle P\nabla Q\rangle = -I(4\pi A/\ell^2)$, and

$$\Delta\phi_H = \frac{\partial}{\partial I}\int_0^{2\pi}\left(-I\frac{4\pi A}{\ell^2}\right)d\theta = -\frac{8\pi^2 A}{\ell^2}$$

as before.

## Exercises

**See also exercises** 7, 8, 9 in Chapter VI.

1. A particle of mass m moves in one dimension x in a potential well

$$V = V_0 \tan^2(\pi x/2a)$$

where $V_0$ and a are constants. Find the action variable I, express the total energy E in terms of I, and find the frequency $\omega = dE/dI$. In particular examine and interpret the low energy ($E \ll V_0$) and high energy ($E \gg V_0$) limits of your expressions (refer to exercise 1.01).

2. A particle of mass m moves in two dimensions (x, y) in a non-isotropic simple harmonic oscillator well

$$V(x,y) = \tfrac{1}{2}m\omega_x^2 x^2 + \tfrac{1}{2}m\omega_y^2 y^2$$

where in general $\omega_x \neq \omega_y$.

---

[12] These variables are obtained from the (s, p) variables by canonical transformation. The generating function $W(s,P,\theta) = \left(\int_0^s ds/r\sin\alpha + \theta\right)P$ is chosen so that the Hamiltonian for the (Q,P) variables does not contain a term proportional to $d\theta/dt$. Since the Hamiltonian for the (s, p) variables has the form $H \approx p^2/2m - pr\sin\alpha\, d\theta/dt$, the function W can be found by solving the Hamilton-Jacobi-like equation $\partial W/\partial\theta = r\sin\alpha\, \partial W/\partial s$.

(a) Find the action variables $(I_x, I_y)$, and express the energy in terms of these.

(b) Find the angle variables $(\phi_x, \phi_y)$, and express the cartesian coordinates in terms of the action-angle variables.

(c) Write down the angle variables and the cartesian coordinates as functions of time.

(d) Sketch the trajectories of the particle in $(x, y)$ space and in $(\phi_x, \phi_y)$ space.

3. A particle of mass m moves in two dimensions $(x, y)$ in a rectangular "infinite square well" potential (sometimes called a rectangular **billiard**)

$$V = 0 \text{ for } 0 < x < a,\ 0 < y < b \quad \text{and} \quad V \to \infty \text{ otherwise}.$$

(a) Find the action variables $I_x$ and $I_y$.

(b) Find the frequencies $\omega_x$ and $\omega_y$, and write down the condition for periodic trajectories. Interpret your result geometrically.

4. A particle of mass m moves in two dimensions $(\rho, \phi)$ in a circular "infinite square well" potential (sometimes called a circular billiard)

$$V = 0 \text{ for } \rho < a \quad \text{and} \quad V \to \infty \text{ for } \rho \geq a.$$

(a) Find the action variables $I_\rho$ and $I_\phi$.

(b) Find the frequencies $\omega_\rho$ and $\omega_\phi$, and write down the condition for periodic trajectories. Interpret your result geometrically.

5. A particle of mass m moves in a three-dimensional isotropic oscillator well

$$V = \tfrac{1}{2}m\omega^2(x^2 + y^2 + z^2) = \tfrac{1}{2}m\omega^2(\rho^2 + z^2) = \tfrac{1}{2}m\omega^2 r^2.$$

(a) Separate the Hamilton-Jacobi equation in cartesian coordinates $(x, y, z)$, find the action variables, and express the Hamiltonian in terms of these. Find the frequencies $(\omega_x, \omega_y, \omega_z)$.

(b) Separate the Hamilton-Jacobi equation in cylindrical coordinates $(\rho, \phi, z)$, find the action variables, and express the Hamiltonian in terms of these. Find the frequencies $(\omega_\rho, \omega_\phi, \omega_z)$.

(c) Separate the Hamilton-Jacobi equation in spherical polar coordinates $(r, \theta, \phi)$, find the action variables, and express the Hamiltonian in terms of these. Find the frequencies $(\omega_r, \omega_\theta, \omega_\phi)$.

6. A particle of mass m moves in a central potential

$$V = -\frac{k}{r} + \frac{h}{r^2}.$$

(a) Find the action variable $I_r$ in terms of the energy E and total angular momentum L.

(b) Use your result to express the energy in terms of the action variables $(I_r, I_\theta, I_\phi)$.

(c) Find the frequencies $(\omega_r, \omega_\theta, \omega_\phi)$. Under what conditions (on the action variables) is the motion periodic?

7. A particle of mass m and charge e moves in a three-dimensional isotropic oscillator well $V = \frac{1}{2}m\omega^2 r^2$, on which is superimposed a uniform magnetic field $\mathbf{B}$. Choosing the symmetric gauge for the vector potential $\mathbf{A} = \frac{1}{2}\mathbf{B} \times \mathbf{r}$ and cylindrical coordinates $(\rho, \phi, z)$ with z-axis in the direction of the magnetic field, show that the time-independent Hamilton-Jacobi equation separates, obtain the action variables $(I_\rho, I_\phi, I_z)$, and express the Hamiltonian in terms of these.

(Ans. $H = (2I_\rho + I_\phi)\sqrt{\omega^2 + \omega_L^2} \mp I_\phi \omega_L + I_z \omega$, where $\omega_L = eB/2mc$ is the Larmor frequency)

8. A simple harmonic oscillator with time-dependent frequency $\omega(t)$ has a Hamiltonian

$$H = \frac{p^2}{2m} + \frac{1}{2}m\omega^2(t)q^2.$$

(a) Transform from $(q,p)$ variables to (instantaneous) action-angle variables $(\phi, I)$. Find, in particular, the Hamiltonian to be used with the action-angle variables.
(b) Write down Hamilton's equations of motion for the action-angle variables.

9. Consider again the simple plane pendulum undergoing small amplitude oscillations, and suppose that the length $\ell$ is shortened adiabatically, this time by pulling the string up through a small hole in the ceiling. Using elementary mechanics, show that the energy of oscillation $E_{osc}$ increases such that $E_{osc}\sqrt{\ell}$ remains constant.

10. A particle of mass m moves in one dimension x between rigid walls at $x = 0$ and at $x = \ell$. Using elementary mechanics:
(a) Show that the average (outward) force on the walls is $F = 2E/\ell$ where E is the (kinetic) energy of the particle.
(b) Suppose now that the wall at $x = \ell$ is moved adiabatically. The energy of the particle then changes as a result of its collisions with the moving wall. Show that $\delta E = -(2E/\ell)\delta\ell$.

(c) Hence show that $E\ell^2$ remains constant under this adiabatic change. Compare this result with that given by "invariance of the action variable."

11. Consider again the Hannay hoop. Write down the Lagrangian (the kinetic energy of the bead in an inertial frame) using as generalized coordinate the displacement s of the bead around the hoop from some fixed point on the hoop. Assume that the hoop is rotating with angular velocity $\Omega = d\theta/dt$. Find the Hamiltonian, and write down Hamilton's equations of motion. Average $(1/\ell)\int_0^\ell \cdots ds$ the right-hand side of these over the position of the bead around the hoop, and integrate with respect to time to obtain the Hannay displacement $\langle \Delta s \rangle = -(2A/\ell)\Delta\theta$ in the

position of the bead. Here A is the total area enclosed by the hoop and $\Delta\theta$ the angle through which the hoop is turned.

(Hint: the Hamiltonian is $H = \frac{p^2}{2m} - pr\sin\alpha\,\Omega - \frac{1}{2}mr^2\cos^2\alpha\,\Omega^2$)

12. Consider the "generalized simple harmonic oscillator" with Hamiltonian

$$H = \tfrac{1}{2}(Xq^2 + 2Yqp + Zp^2)$$

where $(X,Y,Z)$ are parameters with $XZ > Y^2$.

(a) Show that the trajectories in phase space are ellipses and hence find the action variable, showing that it is $I = E/\omega$ where $\omega = \sqrt{XZ - Y^2}$ is the frequency.

(b) Express the variables $(q,p)$ in terms of the action-angle variables $(\phi, I)$. (There are various ways to do this; one way is to solve the Hamilton-Jacobi equation to find the appropriate generating function.)

(c) Suppose that the parameters $\mathbf{R} = (X,Y,Z)$ are changed adiabatically (but always with $XZ > Y^2$) so as to take the system around a closed circuit in parameter space. Show that the resulting Hannay angle is

$$\Delta\phi_H = \iint \frac{\mathbf{R}\cdot d\mathbf{S}}{4\omega^3}.$$

(J. H. Hannay, "Angle variable holonomy in adiabatic excursion of an integrable Hamiltonian," J. Phys. A **18**, 221-230 (1985); M. V. Berry, "Classical adiabatic angles and quantal adiabatic phase," J. Phys. A **18**, 15-27 (1985).)

# CHAPTER X

# NON-INTEGRABLE SYSTEMS

We have been concerned mainly with **integrable** systems. Such systems are characterized by the existence of a set of f independent analytic single-valued constants of the motion, such as (or which may be taken to be) the f action variables $I_a$. Further, these constants of the motion must be in **involution** with one another; their Poisson brackets with one another must vanish, $[I_a, I_b] = 0$. Integrable systems, while physically important, are in fact rare and non-generic, so let us consider briefly in this last chapter the much more difficult non-integrable systems. Those who wish to pursue these matters more deeply may consult the excellent review articles of Berry,[1] Helleman,[2] and Hénon,[3] and the texts of Arnol'd,[4] Gutzwiller,[5] and Lichtenberg and Lieberman.[6]

## Surface of section

We consider for the most part conservative systems, systems with time-independent Hamiltonians. Such systems in one degree of freedom are always integrable; they have one single-valued constant of the motion (e.g. the Hamiltonian). The simplest such non-integrable system, which is the case we focus on here, thus has two degrees of freedom. Phase space $(q_1, p_1; q_2, p_2)$ is then four-dimensional, but for conservative systems the system point is restricted to a three-dimensional constant energy "surface" $H(q_1, p_1; q_2, p_2) = E$ in phase space. Thus, one way to depict the motions at given energy is to draw the trajectories in the three-dimensional space $(q_1, p_1; q_2)$ with $p_2$ determined by the energy (Fig. 10.01), but this is clearly inconvenient. Instead, we shall see that we can learn a great deal about a system from the simpler option of studying its behavior on a two-dimensional slice through the constant energy "surface." This slice is called a **Poincaré surface of section**. It can be obtained by setting, for example, $q_2 = 0$. The

---

[1]M. V. Berry, "Regular and Irregular Motion," in *Topics in Nonlinear Dynamics, A Tribute to Sir Edward Bullard*, AIP Conference Proceedings, No. 46 (American Institute of Physics, New York, 1978), ed. Siebe Jorna, pp. 16-120.

[2]Robert H. G. Helleman, "Self-Generated Chaotic Behavior in Nonlinear Mechanics," in *Fundamental Problems in Statistical Mechanics* (North-Holland Publishing Company, Amsterdam, 1980), ed. E. G. D. Cohen, pp. 165-233.

[3]Michel Hénon, "Numerical Exploration of Hamiltonian Systems," in *Les Houches Session XXXVI, 1981, Chaotic Behaviour of Deterministic Systems* (North-Holland Publishing Company, Amsterdam, 1983), eds. Gerard Iooss, Robert H. G. Helleman and Raymond Stora, pp. 53-170.

[4]V. I. Arnol'd, *Mathematical Methods of Classical Mechanics* (Springer-Verlag Inc., New York, 1978), trans. K. Vogtmann and A. Weinstein.

[5]Martin C. Gutzwiller, *Chaos in Classical and Quantum Mechanics* (Springer-Verlag Inc., New York, 1990).

[6]A. J. Lichtenberg and M. A. Lieberman, *Regular and Chaotic Dynamics*, 2nd ed. (Springer-Verlag Inc., New York, 1992).

remaining pair of conjugate variables $(q_1, p_1)$ can then serve as coordinates on the surface of section.

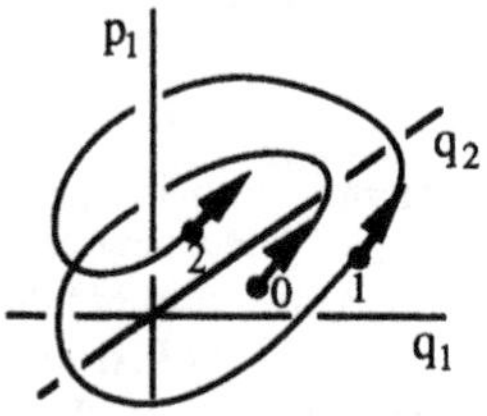


Fig. 10.01. Trajectory in 3D constant energy "surface"

If we start the system at some point (Fig. 10.01, point 0) on the surface of section, it normally wanders off into the third dimension on what may be a long and complicated trajectory. For bounded motion, however, it eventually cuts through the surface of section again in the original direction (Fig. 10.01, point 1). (Between two such cuttings is a cutting in the opposite direction.) This is repeated over and over, so for a given start (that is, a given trajectory) we obtain a sequence of points on the surface of section marking the successive cuttings of the surface by the trajectory of the system (Fig. 10.01, points 0, 1, 2, ⋯). Different starts usually give different sequences, so by trying a variety of starts we can build up a variety of patterns of points on the surface.

Instead of following a given trajectory as it cuts successively through the surface of section, we can consider this instead as a **mapping** of the surface of section onto itself,

$$X \to X' = F(X),$$

where X stands for the pair $(q,p)$. We can find the relation between X and $X'$, the mapping, by integrating the equations of motion, numerically if necessary, to obtain the trajectory from a given start point X on the surface of section to the mapped point $X'$, the first cutting in the original direction of the surface of section by the trajectory. Mappings can, of course, arise in other ways. For example, in studying the general behavior of dynamical systems we may for simplicity and speed in computing replace a system with continuous time development by a model system with discrete time development.

For Hamiltonian systems this mapping has an important property which we can discover by considering the trajectories that start in a little patch of area on the surface of section. These form a tube which eventually cuts again through the surface of section in the original direction; the old patch of area $\omega_\Sigma$ is thereby mapped into a new patch $\omega'_\Sigma$ (Fig. 10.02).

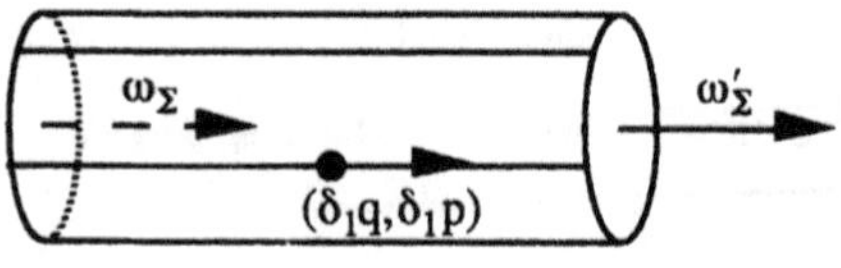


Fig. 10.02. Tube of trajectories

Hamiltonian dynamics is such that the area of the patch is invariant under the mapping, so $\omega'_\Sigma = \omega_\Sigma$. To show this, note that the integral of the canonical two-form $\omega = \delta_1 q \delta_2 p - \delta_1 p \delta_2 q$ over any closed surface, and in particular over the surface of the tube of trajectories, ends plus side, is zero. Over the new end it is $\omega'_\Sigma$, and over the old end it is $-\omega_\Sigma$. The side of the tube is composed of trajectories, so here we can always take one of the displacements of the two-form in the direction of the trajectories,

$$\delta_1 q = \varepsilon \frac{\partial H}{\partial p} \qquad \delta_1 p = -\varepsilon \frac{\partial H}{\partial q}.$$

The two-form on the side of the tube then becomes

$$\omega = \varepsilon \frac{\partial H}{\partial p} \delta_2 p + \varepsilon \frac{\partial H}{\partial q} \delta_2 q = \varepsilon \delta_2 H = 0,$$

since the trajectories are all at the same energy. Thus the side of the tube gives zero contribution to the integral of the two-form, and we have $\omega'_\Sigma - \omega_\Sigma + 0 = 0$ which is what we wanted to show. It should be emphasized that $\omega'_\Sigma$ is *not* $\omega_\Sigma$ at some later time. System points which start in $\omega_\Sigma$ at the same time arrive back at the surface of section at slightly differing times, so the original patch is "tipped" when it returns to the vicinity of the surface of section. The results of Chapter VII, which show that the area of this tipped patch is equal to the area of the original patch, are thus not relevant here.

Of special importance on the surface of section are the **fixed points** $X_0$,

$$X_0 = F(X_0),$$

which remain unchanged under the mapping ($X_0 \to X_0$), and the finite **cycles** of points around which the system steps on successive applications of the mapping ($X_0 \to X_1 \to \cdots \to X_0$). A cycle as a whole is invariant under the mapping, and if the cycle has s distinct points, each point is a fixed point of the mapping $F^s$. These fixed points and cycles correspond to the periodic orbits, those closed orbits of the system which eventually return to their start.

What happens *near* a fixed point $X_0$? We specify points near $X_0$ by the *small* relative coordinate x. Then to first order in small quantities the mapping is *linear*

$$x' = Mx$$

where M is a two by two matrix with constant coefficients. Further, since the mapping is area-preserving

$$\det M = 1.$$

Now the way to handle this sort of problem is well known from studies of coupled oscillators or from quantum mechanics. We begin by finding the eigenvalues $\lambda$ and eigenvectors of M. This is useful because under the mapping the eigenvectors behave simply: they get multiplied by $\lambda$. The original variables can then be found as linear combinations of the eigenvectors. The behavior depends on the nature of the eigenvalues. As we can easily show, the eigenvalues satisfy the quadratic equation

$$\lambda + 1/\lambda = \operatorname{Tr} M$$

where $\operatorname{Tr} M$ is the trace of the matrix M. This form of the equation brings out the fact that the roots are reciprocals of one another. There are two types of root, depending on whether the discriminant is negative or positive. If $|\operatorname{Tr} M| < 2$, the discriminant is negative and the two roots are complex numbers; indeed, they are complex conjugates of one another. This, combined with the fact that they are reciprocals, shows that they have the form $\lambda = e^{\pm i\alpha}$, with $2\cos\alpha = \operatorname{Tr} M$. If $|\operatorname{Tr} M| > 2$, the discriminant is positive and the two roots are real. There are then two subcases: (a) $\operatorname{Tr} M > 2$, which leads to positive roots $\lambda = e^{\pm\beta}$, one greater and one less than 1, and (b) $\operatorname{Tr} M < -2$, which leads to negative roots $\lambda = -e^{\pm\beta}$, in both cases with $2\cosh\beta = |\operatorname{Tr} M|$.

First consider the case $|\operatorname{Tr} M| < 2$. The (complex) eigenvectors $e_\pm$ satisfy the eigenvalue equation

$$M e_\pm = e^{\pm i\alpha} e_\pm .$$

Any start $x(0)$ can be written as a linear combination of these eigenvectors,

$$x(0) = a_+ e_+ + a_- e_- .$$

After n intersections of the trajectory with the surface of section (in a given direction), the start is mapped into

$$x(n) = M^n x(0) = e^{in\alpha} a_+ e_+ + e^{-in\alpha} a_- e_- .$$

Now $x(0)$ and $x(n)$ are real, so $a_+ e_+$ and $a_- e_-$ are complex conjugates of one another and we can set

$$2 a_\pm e_\pm = a_1 e_1 \pm i a_2 e_2$$

where $a_1 e_1$ and $a_2 e_2$ are real vectors. Thus at "time" n the system is at

$$x(n) = a_1 e_1 \cos n\alpha - a_2 e_2 \sin n\alpha .$$

The "motion" is simple harmonic in the $e_1$ direction, and simple harmonic $\pi/2$ out of phase in the $e_2$ direction. The resulting sequence of points thus lies on an ellipse (Fig. 10.03), and the fixed point on which the ellipse is centered is called an **elliptic fixed point**.

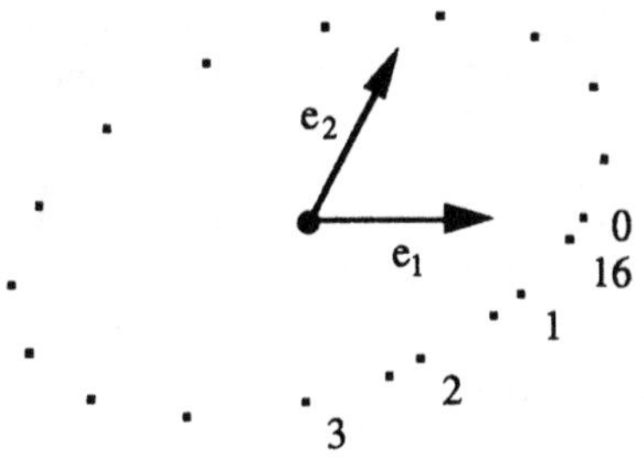


Fig. 10.03. Trajectory near an elliptic fixed point

Now consider the case $\mathrm{Tr}\,M > 2$. In this case the eigenvectors are real and define four directions $\pm e_+$ and $\pm e_-$ in the surface of section. If the start $x(0)$ is in one of these directions, each successive mapping brings the system point a factor $e^{-\beta}$ nearer to the fixed point for the $e_+$ directions, and a factor $e^{+\beta}$ further from the fixed point for the $e_-$ directions, with the displacements occurring along the initial directions in both cases. The $\pm e_+$ and $\pm e_-$ directions are the ends, near the fixed point, of the separatrices. In general we write the start $x(0)$ as a linear combination of the vectors $e_+$ and $e_-$ and after n mappings find the system at

$$x(n) = e^{-n\beta}\, a_+ e_+ + e^{n\beta}\, a_- e_- .$$

These points lie on a hyperbola (Fig. 10.04(a)) so this type of fixed point is called a **hyperbolic fixed point**. The case $\mathrm{Tr}\,M < -2$, with negative eigenvalues, is similar, except that instead of staying on one branch of the hyperbola, the successive points jump back and forth between two branches (Fig. 10.04(b)).

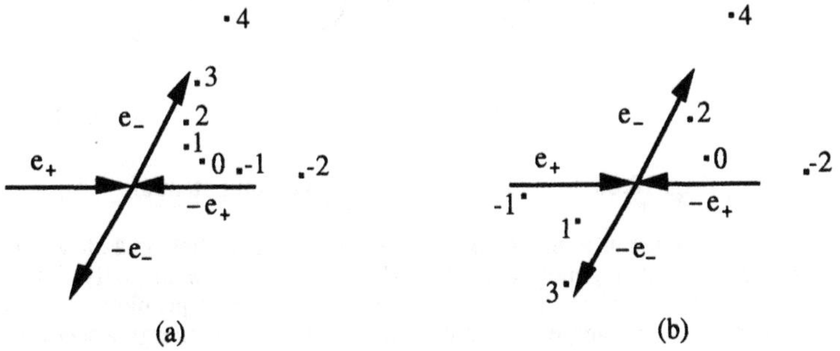


Fig. 10.04. Trajectory near a hyperbolic fixed point

There is actually a third type of fixed point, intermediate between the elliptic and hyperbolic fixed points, called a **parabolic fixed point.** It occurs if $|\mathrm{Tr}M| = 2$, so the discriminant is zero. If $\mathrm{Tr}M = +2$, the single eigenvalue is $\lambda = +1$. The single real eigenvector e is invariant, $Me = e$, under the mapping, and its direction gives a *line* of fixed points which passes through the original point. Points not on the line are moved under the mapping parallel to the line an amount proportional to their displacement from the line, so the mapping is a pure shear. If $\mathrm{Tr}M = -2$, the eigenvalue is $\lambda = -1$, and the eigenvector e undergoes inversion, $Me = -e$, under the mapping. The mapping is a pure shear $(-M)$ together with an inversion $(-1)$.

## Integrable and non-integrable systems

Let us now look at typical surfaces of section for integrable and for non-integrable systems.

We first consider integrable systems. As an example we take the bound central force problem in two dimensions. The Hamiltonian (in cartesian coordinates) is

$$H = \frac{1}{2m}\left(p_x^2 + p_y^2\right) + V\left(\sqrt{x^2 + y^2}\right).$$

The surface of section can be taken as $y = 0$ with $p_y > 0$. Coordinates on the surface are then $(x, p_x)$ (Fig. 10.05).

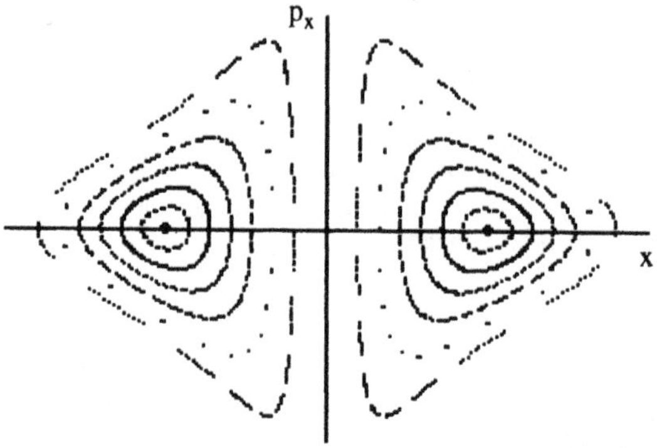


Fig. 10.05. Surface of section for a central potential

We find that the sequence of points resulting from a given start lies on a closed loop. For most starts and for non-degenerate central force the points eventually fill this loop densely. The reason is that the two-dimensional central force problem, or indeed any integrable system with two degrees of freedom, has besides the energy a second constant of the motion. For the central force problem it is the angular momentum

$$L = xp_y - yp_x.$$

The variables on the surface of section $y = 0$ are then restricted by the two conditions

$$\frac{1}{2m}\left(p_x^2 + p_y^2\right) + V(|x|) = E \quad \text{and} \quad xp_y = L$$

with $p_y > 0$. Eliminating $p_y$ between these, we see that for given energy E and angular momentum L the coordinates $(x, p_x)$ on the surface of section are related by

$$\frac{p_x^2}{2m} = E - \frac{L^2}{2mx^2} - V(|x|)$$

where x has the same sign as L. This relation is the same as that between position and momentum for one-dimensional motion in a potential $V_{eff}(x) = L^2/2mx^2 + V(|x|)$, for which the phase space trajectories[7] for bound motions are closed loops. These are the loops on which the sequences of points on the surface of section lie. Occasionally one chooses a start which is on a periodic orbit. In this case the points on the surface of section form a finite repeating set, a cycle, and do not fill a loop. Note for example the central point, an elliptic fixed point, or the second largest "loop" in Fig. 10.05. If the system is completely degenerate, as for a particle moving in a gravitational potential, all orbits are periodic and all starts give a finite cycle. In this case there is yet another constant of the motion. For the gravitational potential it is the direction of pericenter (the direction of the Laplace-Runge-Lenz vector).

We now consider a system (which turns out to be non-integrable) first studied by **Hénon and Heiles**:[8] a particle of unit mass moves in the two-dimensional potential well

$$V = \tfrac{1}{2}(x^2 + y^2) + \tfrac{1}{3}(3y^2x - x^3).$$

This may be thought of as an isotropic harmonic oscillator potential perturbed by cubic terms. These have been chosen so that the potential has three-fold symmetry, as can be seen by writing the potential in plane polar coordinates, $V = \frac{1}{2}r^2 - \frac{1}{3}r^3\cos 3\theta$. The potential is zero at the origin and initially increases as we move away from the origin, reaching the constant value $V = \frac{1}{6}$ on the sides of the equilateral triangle which has vertices at $(1,0)$, $\left(-\frac{1}{2}, \frac{\sqrt{3}}{2}\right)$, and $\left(-\frac{1}{2}, -\frac{\sqrt{3}}{2}\right)$. Outside the triangle the potential tends to $-\infty$ as $r \to \infty$ in the three angular sectors of "width" $\pi/3$ centered on the directions of the vertices, and it tends to $+\infty$ in the other three sectors (Fig. 10.06).

---

[7]For fixed E and for various values of L; that is, for various starts of given energy.

[8]M. Hénon and C. Heiles, "The Applicability of the Third Integral of Motion: Some Numerical Experiments," Astron. J. **69**, 73-79 (1964); F. G. Gustavson, "On Constructing Formal Integrals of a Hamiltonian System Near an Equilibrium Point," Astron. J. **71**, 670-686 (1966).

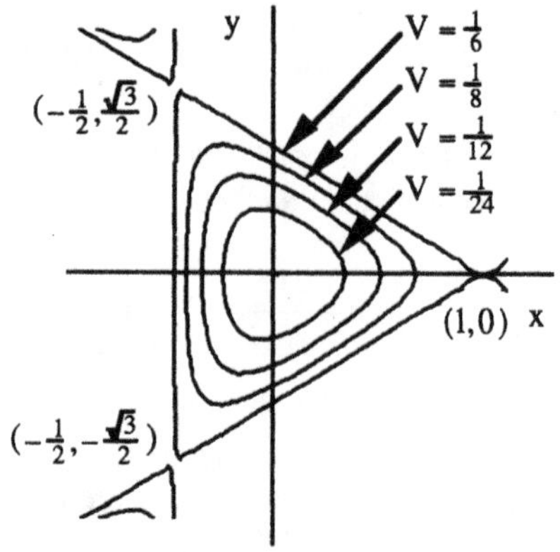


Fig. 10.06. Hénon-Heiles potential

Motion outside the triangle or with too high an energy is unbounded, but if the particle starts in the triangular region and with total energy $E < \frac{1}{6}$, it remains forever trapped in this region. Further, the momenta are limited by $\frac{1}{2}(p_x^2 + p_y^2) < \frac{1}{6}$. The system motion is then bounded. While we cannot solve this problem analytically (no second constant of the motion is known), the problem does not seem complicated or unusual, and without prior experience we probably would not expect anything very much different from the previous (integrable) case to happen. Let us see what a surface of section study, as carried out by Hénon and Heiles, reveals about the motion. We start the system at some point on the surface of section $y = 0$ going from front to back and with energy $E < \frac{1}{6}$ and integrate the equations of motion numerically, following the orbit as it cuts again and again through the surface going from front to back. The resulting sequences of points for various starts and for energies $E = \frac{1}{12}, \frac{1}{8}$, and $\frac{1}{6}$ are shown in Fig. 10.07.

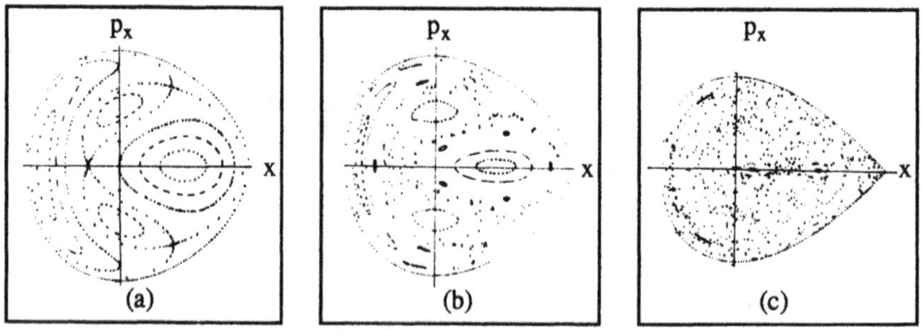


Fig. 10.07. Surface of section for the Hénon-Heiles potential at energy (a) $E = \frac{1}{12}$; (b) $E = \frac{1}{8}$; (c) $E = \frac{1}{6}$

At the lowest energy $E = \frac{1}{12}$ (Fig. 10.07(a)) the regular lines of points surround what appear to be four "elliptic points." Also shown are what appear to be "separatrices" joining three "hyperbolic points." These results suggest that there may be a second constant of the motion which we have been unable to find. Note, however, the slight fuzz near the "hyperbolic points," which is a foretaste of things to come. An increase in energy to $E = \frac{1}{8}$ changes the picture entirely (Fig. 10.07(b)). While we can still see the four "elliptic points" surrounded by closed "curves" (but notice that one of the "curves" appears to have broken up into five "elliptic points" and their surrounding "curves"), the former "separatrices" appear to have degenerated into an irregular jumble of points. What is even more surprising is that all these isolated points are the result of *one* trajectory cutting through the surface of section. It is unlikely that we can pass a simple curve through these points. If the energy is increased still further to $E = \frac{1}{6}$ (Fig. 10.07(c)) the regular regions surrounding the "elliptic points" have practically disappeared and most of the accessible region $\frac{1}{2}p_x^2 + \frac{1}{2}x^2 - \frac{1}{3}x^3 \leq \frac{1}{6}$ is filled by an irregular collection of points, all from a single trajectory. The largely regular motion of Fig. 10.07(a) has become the largely irregular motion of Fig. 10.07(c).

It must be emphasized that the results for the Hénon-Heiles system, and not those for the central potential, are the normal "generic" results for a conservative mechanical system with two degrees of freedom. Pick a system "at random," and it almost certainly behaves like the Hénon-Heiles system. Our job is now to try to understand how and why the various features which we have seen in the Hénon-Heiles system occur.

## Perturbation theory

Suppose we start with an integrable system, for which we can introduce action-angle variables $(\phi, I)$. The motions then occur on tori in phase space, a particular torus being labeled by the action variables I, and points on the torus being labeled by the angle variables $\phi$. The Hamiltonian has the form $H_0(I)$, a function of the action variables alone. Now suppose we change the Hamiltonian slightly by adding a small perturbation $\varepsilon H_1(\phi, I)$. The new Hamiltonian is

$$H(\phi, I) = H_0(I) + \varepsilon H_1(\phi, I).$$

The variables $(\phi, I)$ are still canonical variables for this new system, but they are no longer action-angle variables; the new Hamiltonian depends on the coordinates $\phi$ as well as on the momenta I. Is this new system integrable? Does a new set of action-angle variables $(\phi', I')$ with associated tori exist, such that the new Hamiltonian depends only on the action variables $I'$? If so, we expect them to differ from the old variables $(\phi, I)$ by an infinitesimal canonical transformation

$$\phi' = \phi + \varepsilon \frac{\partial G(\phi, I')}{\partial I'} \qquad I = I' + \varepsilon \frac{\partial G(\phi, I')}{\partial \phi}$$

with some generator G to be determined. The new Hamiltonian $H'(I')$ is to be a function of the new action variables $I'$ alone. Now $H'$ is obtained simply by changing variables in $H(\phi,I)$, so we have

$$H'(I') = H_0\left(I' + \varepsilon\frac{\partial G(\phi,I')}{\partial\phi}\right) + \varepsilon H_1\left(\phi, I' + \varepsilon\frac{\partial G(\phi,I')}{\partial\phi}\right)$$

$$= H_0(I') + \varepsilon\left(\boldsymbol{\omega}_0(I')\cdot\frac{\partial G(\phi,I')}{\partial\boldsymbol{\phi}} + H_1(\phi,I')\right) + \cdots$$

where the $\boldsymbol{\omega}_0 = \partial H_0/\partial\mathbf{I}'$ are the unperturbed frequencies. To see the implications of this, we expand the generator G and the perturbing Hamiltonian $H_1$ in multiple Fourier series in the old angle variables,

$$G(\phi,I') = \sum_{\mathbf{n}} g_{\mathbf{n}}(I')e^{i\mathbf{n}\cdot\boldsymbol{\phi}} \quad \text{and} \quad H_1(\phi,I') = \sum_{\mathbf{n}} h_{1\mathbf{n}}(I')e^{i\mathbf{n}\cdot\boldsymbol{\phi}}.$$

The expansion of the new Hamiltonian becomes

$$H'(I') = H_0(I') + \varepsilon\left(h_{1\mathbf{n}=0}(I') + \sum_{\mathbf{n}\neq 0}\left(i\mathbf{n}\cdot\boldsymbol{\omega}_0(I')g_{\mathbf{n}}(I') + h_{1\mathbf{n}}(I')\right)e^{i\mathbf{n}\cdot\boldsymbol{\phi}}\right) + \cdots.$$

The $\mathbf{n} = 0$ term gives the new Hamiltonian

$$H'(I') = H_0(I') + \varepsilon h_{1\mathbf{n}=0}(I') + \cdots = H_0(I') + \varepsilon\langle H_1\rangle + \cdots,$$

where we have used the fact that $h_{1\mathbf{n}=0}(I')$ is equal to the average $\langle H_1(\phi,I')\rangle$ of the perturbing Hamiltonian over the angle variables. The $\mathbf{n}\neq 0$ terms give the Fourier components of the appropriate generator,

$$g_{\mathbf{n}}(I') = \frac{ih_{1\mathbf{n}}(I')}{\mathbf{n}\cdot\boldsymbol{\omega}_0(I')},$$

so the generator is[9]

$$G(\phi,I') = \sum_{\mathbf{n}\neq 0}\frac{ih_{1\mathbf{n}}(I')}{\mathbf{n}\cdot\boldsymbol{\omega}_0(I')}e^{i\mathbf{n}\cdot\boldsymbol{\phi}}.$$

---

[9]The (angle-independent) n=0 term in G remains undetermined. All such a term does is shift the "zero points" of the angle variables.

At first this would appear to provide the required modification to first order in $\varepsilon$ to the action-angle variables. Indeed, these results are often useful.[10] On closer examination, however, there is a problem. For rational tori, those for which the ratio $\omega_1/\omega_2$ of the frequencies is a rational number $r/s$, the denominator $\mathbf{n}\cdot\boldsymbol{\omega}_0$ in G can be zero. Even for irrational tori the denominator can become, for suitable sufficiently large $\mathbf{n}$, arbitrarily small, so the convergence of the series in $\mathbf{n}$ is highly suspect. This is the famous **problem of small denominators**. Apart from this there is also the problem of extending the calculation to higher orders in $\varepsilon$, and of proving the convergence of that process. For these questions ordinary perturbation theory does not work very well, if at all.

## Irrational tori

The solution to these problems was finally suggested by Kolmogorov (1954) and proved by Arnol'd (1963) and Moser (1962), who developed an approximation scheme with vastly improved convergence. With its aid Kolmogorov, Arnol'd, and Moser were able to show that "sufficiently irrational" tori, those for which the frequency ratio $\omega_1/\omega_2$ is sufficiently far from a rational number, may be distorted by a weak perturbation but retain their structure. Such tori continue to exist after perturbation; they are stable. This is the famous **KAM theorem**. We shall not attempt to go into these ideas; the essence of the scheme can be found in the references in the introduction to this chapter.

It is interesting, however, to look into what is meant by "sufficiently irrational." To understand this, we must consider how to approximate an irrational number such as $\pi$ by rationals. One way is to truncate the decimal expansion $\pi = 3.14159\cdots$. This leads to the sequence of decimal approximates

$$\pi = \frac{3}{1}, \frac{31}{10}, \frac{314}{100}, \frac{3142}{1000}, \cdots.$$

---

[10] For example, they can be used to obtain the correction to the frequency of a simple pendulum for finite amplitude oscillations. The potential energy is

$$V = mg\ell(1-\cos\theta) \approx \tfrac{1}{2}mg\ell\theta^2 - \tfrac{1}{24}mg\ell\theta^4 + \cdots.$$

The second order term in $\theta$ leads to simple harmonic motion at frequency $\omega_0 = \sqrt{g/\ell}$. The fourth and higher order terms modify this result. Let us treat the fourth order term as a perturbation. We express it in terms of the unperturbed action-angle variables $(\phi, I)$ by setting $\theta = \sqrt{2I/m\ell^2\omega_0}\,\sin\phi$. The perturbation is then

$$H_1 = -\frac{1}{24}mg\ell\theta^4 = -\frac{I^2}{6m\ell^2}\sin^4\phi.$$

Its average over a cycle,

$$\langle H_1\rangle = -\frac{I^2}{6m\ell^2}\langle\sin^4\phi\rangle = -\frac{I^2}{16m\ell^2},$$

gives the correction to the Hamiltonian. Differentiating with respect to the action I then gives the correction $\omega_1$ to the frequency,

$$\omega_1 = \frac{\partial\langle H_1\rangle}{\partial I} = -\frac{I}{8m\ell^2} = -\frac{1}{16}\omega_0\theta_0{}^2$$

where $\theta_0$ is the amplitude of the oscillation.

If we write one of these as r/s, the error is less than 1/s. But we can do much better than this by using a continued fraction

$$a_0 + \cfrac{1}{a_1 + \cfrac{1}{a_2 + \cdots}}$$

To find the continued fraction representation of a number, we subtract the integer part ($a_0$) of the number, invert and subtract the integer part ($a_1$), invert and subtract the integer part ($a_2$), invert and .... For $\pi$ this gives

$$\pi = 3 + \cfrac{1}{7 + \cfrac{1}{15 + \cfrac{1}{1 + \cfrac{1}{292 + \cdots}}}}$$

Truncating at the n'th stage gives a rational approximate. For $\pi$ we have the sequence of continued fraction approximates

$$\pi = \frac{3}{1}, \frac{22}{7}, \frac{333}{106}, \frac{355}{113}, \ldots.$$

We can show that these differ from $\pi$ by less than $1/s^2$,

$$\left|\pi - \frac{r}{s}\right| < \frac{1}{s^2},$$

and are the best rational approximates there are, in the sense that no rational fraction with a smaller denominator does better.

Now KAM requires that the frequency ratio $\omega_1/\omega_2$ be further away from any rational number r/s than $k(\varepsilon)/s^{2.5}$,

$$\left|\frac{\omega_1}{\omega_2} - \frac{r}{s}\right| \geq \frac{k(\varepsilon)}{s^{2.5}}.$$

The function $k(\varepsilon)$ is unspecified but goes to zero with the perturbation strength $\varepsilon$. The excluded regions around the first few lowest order fractions in the interval 0 to 1 are shown in Fig. 10.08.

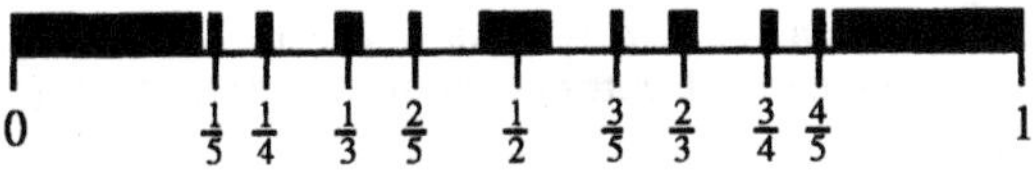


Fig. 10.08. Some frequency ratio regions excluded by KAM

It is important to realize that after taking out these regions there is still something left. We can estimate the total length of the excluded region by multiplying the length $2k(\varepsilon)/s^{2.5}$ around a fraction with denominator s by the number s of such fractions and summing,

$$L < \sum_{s=1}^{\infty} \frac{2k(\varepsilon)}{s^{2.5}} \times s = 2k(\varepsilon) \sum_{s=1}^{\infty} \frac{1}{s^{1.5}} = 5.224 \times k(\varepsilon).$$

This is an *over*estimate of the excluded length since many regions overlap. We see that for weak perturbation the factor $k(\varepsilon)$ and the excluded length tend to zero, and thus the frequency ratio for *most* tori is "sufficiently irrational." So for weak perturbation *most* tori continue to exist, and we expect most of the surface of section to be covered by regular lines of points, much like an integrable system. Between these lines, however, are the former rational tori, and the regions near them increase with the perturbation strength. As we see in the next section, these rational tori break up in a complex way, and it is this that gives the surface of section for a non-integrable system its character.

## Rational tori

In order to see what happens to a rational torus, we first return to an integrable system in two degrees of freedom, looking at it this time from the point of view of the action-angle variables $(\phi_1, I_1; \phi_2, I_2)$. The action variables $I_1$ and $I_2$ for a given motion are constant; this defines the two-dimensional torus. The energy $E = H(I_1, I_2)$ is constant on a given torus. Position on the torus is given by the angle variables $\phi_1$ and $\phi_2$; these increase uniformly with time,

$$\phi_1 = \phi_1(0) + \omega_1 t \qquad \phi_2 = \phi_2(0) + \omega_2 t.$$

The frequencies are given by $\omega_1 = \partial H/\partial I_1$ and $\omega_2 = \partial H/\partial I_2$ and are, in general, functions of the actions.

We take the surface of section $\phi_2 = 0$. Suitable coordinates on the surface are then $(\phi_1, I_1)$. It is convenient to regard these, or rather $\phi_1$ (angle) and $\sqrt{2I_1}$ (radius), as plane polar coordinates (Fig. 10.09).[11] We start the system at some point $(\phi_1(0), I_1(0))$ on the surface of section $(\phi_2 = 0, I_2)$, where $I_2$ is determined by the values of $I_1$ and E. In a time $2\pi/\omega_2$ the angle variable $\phi_2$ increases by $2\pi$ and the trajectory again cuts through the surface in the original direction. In this time the angle variable $\phi_1$ increases

[11]The element of area on the surface is then $\sqrt{2I}\, d\sqrt{2I}\, d\phi = dI\, d\phi$.

by $2\pi\omega_1/\omega_2$ while the action variable $I_1$ remains constant. This is repeated over and over. If we label the sequence of cuttings by an integer n, then (now dropping the subscript 1)

$$\phi(n+1) = \phi(n) + 2\pi\alpha \qquad I(n+1) = I(n)$$

where $\alpha(I) = \omega_1/\omega_2$ is called the **rotation number**. It is the fraction of a rotation by which $\phi$ increases per cutting. It usually depends on the action variable I.

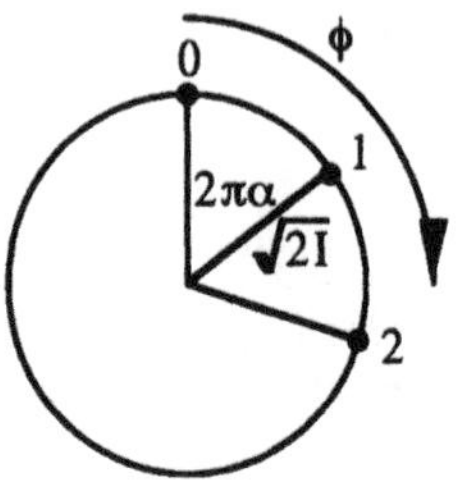


Fig. 10.09. Surface of section in action-angle variables

The sequence of points lies on a circle of radius $\sqrt{2I}$, with successive points separated in angle by $2\pi\alpha$. The nature of this sequence depends on the rotation number. If $\alpha$ is an irrational number, the sequence of points never repeats and eventually covers the circle densely. If, however, $\alpha$ is a rational number, say $\alpha = r/s$, then after s cycles $\phi$ increases by $2\pi r$ and the system point is back to its start. We can view the above as a mapping, called the **twist mapping**, of the surface of section onto itself. In this mapping circles of constant I are rotated, but the amount of rotation depends on the radius of the circle so radial lines become curved. For rational circles the start is a fixed point of the $(\text{mapping})^s$.[12] Indeed, since the rotation number is independent of angle, *every* point of a rational circle is a fixed point of the $(\text{mapping})^s$.

So far we have merely restated what we already know about integrable systems. Now suppose we apply a weak perturbation to this system. What happens to the circles depends on whether the circle is an irrational or a rational one. According to KAM the irrational circles are slightly distorted but they essentially retain their structure. The behavior of the rational circles under perturbation is much more complicated. As we might expect, most of the circle of fixed points is destroyed, but one can show that an even number 2ks $(k = 1,2,\cdots)$ of fixed points remains. This is the **Poincaré-Birkhoff theorem**. Further, half of these fixed points are of elliptic type and half are of hyperbolic type. Each of these new elliptic points is a center surrounded by its closed loops. On most of these secondary loops the frequency ratio is irrational, but for some it is rational, and these in turn break up into an even number of fixed points, half elliptic and half hyperbolic. This picture is repeated over and over, on finer and finer scales, ad infinitum (Fig. 10.10).

---

[12] What *type* of fixed point is it?

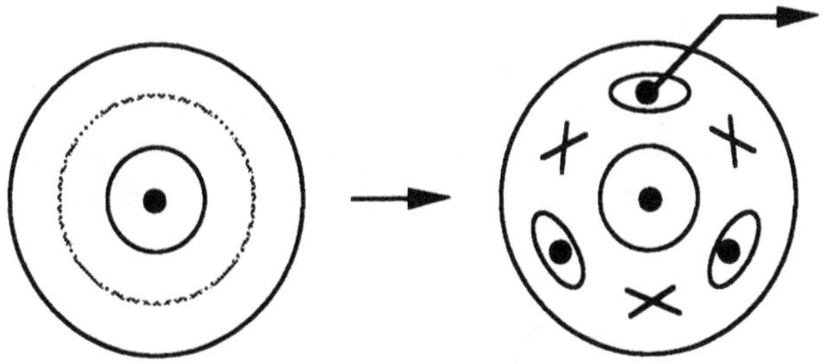

Fig. 10.10. Irrational circles (solid) and rational circle (dotted) under perturbation, first step

But this is not all! We still have to consider the behavior near the hyperbolic points. Near such points are the ends of four invariant curves, the separatrices, each of which maps onto itself (Fig. 10.11). On two of these ($\Gamma_+$) successive points approach closer and closer to the hyperbolic point without ever reaching it, and on two ($\Gamma_-$) they recede further and further away.

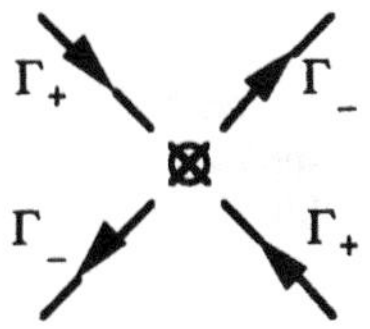


Fig. 10.11. Invariant curves near a hyperbolic point

Let us see what happens as we move away from the hyperbolic point on one of the $\Gamma_+$ curves. For an integrable system, such as a simple pendulum, the $\Gamma_+$ curve eventually joins smoothly one of the $\Gamma_-$ curves from the same or a different hyperbolic point (Fig. 10.12(a)). For a non-integrable system, however, the $\Gamma_+$ and $\Gamma_-$ curves do not join smoothly but intersect (Fig. 10.12(b)). The point of intersection is called a **homoclinic point** (or heteroclinic if the initial hyperbolic points are different). The existence of such points and the consequences which we are about to discuss were first realized by Poincaré.

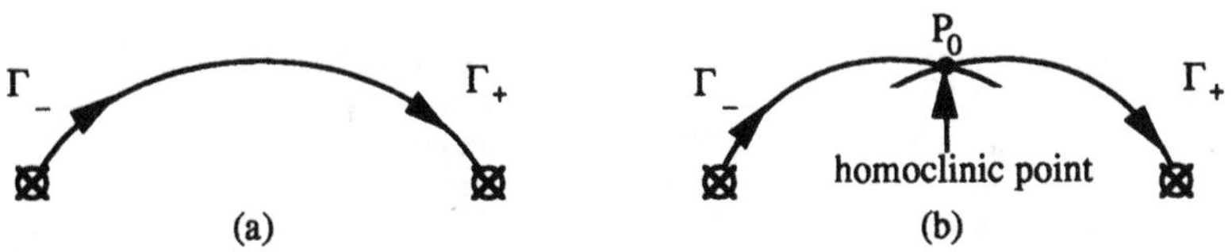


Fig. 10.12. Invariant curves for (a) an integrable system (b) a non-integrable system

Consider successive mappings of the homoclinic point. These are on the $\Gamma_+$ curve and hence approach closer and closer to the hyperbolic point, with the distance along the $\Gamma_+$ curve between two successive ones decreasing exponentially (Fig. 10.13(a)). On the other hand, they are also on the $\Gamma_-$ curve and hence recede further and further from the hyperbolic point, with the distance along the $\Gamma_-$ curve between two successive ones increasing exponentially. To accomplish this, the $\Gamma_-$ curve must weave back and forth across the $\Gamma_+$ curve, with the loops getting longer and longer and thinner and thinner (Fig. 10.13(b)). (The direction of crossing is preserved so between any two mappings of the homoclinic point is a crossing in the opposite direction.)

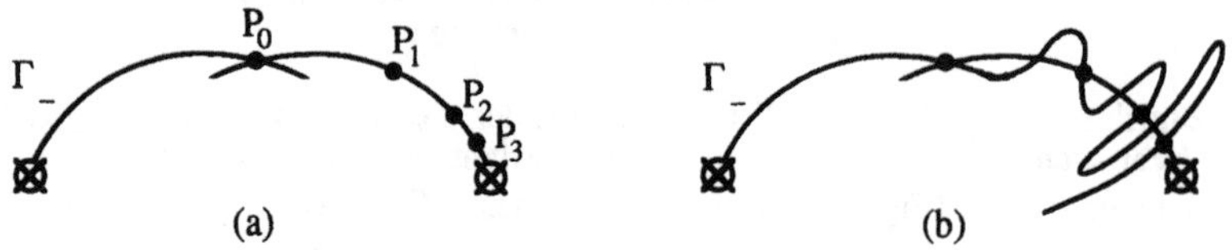


Fig 10.13. (a) Successive mappings of point $P_0$
(b) Continuation of curve $\Gamma_-$

A similar thing happens if we consider the pre-images of the initial homoclinic point, obtained by applying successively the time reversed or inverse mapping; only the roles of the $\Gamma_+$ and $\Gamma_-$ curves are reversed. Also, similar considerations apply to the other pair of $\Gamma_+$ and $\Gamma_-$ curves associated with the hyperbolic point. The resulting tangle is shown in Fig. 10.14.

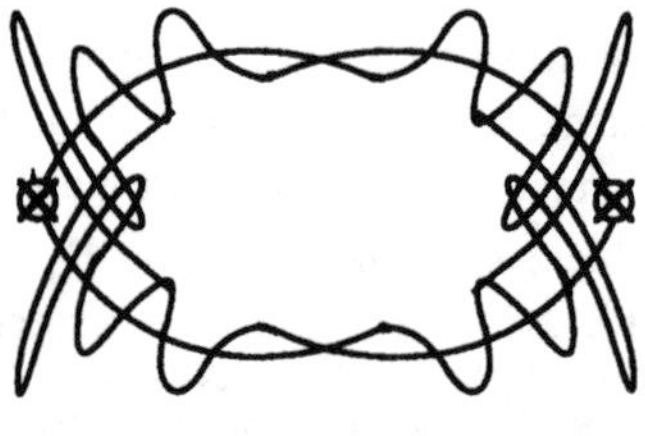

Fig. 10.14. Separatrices $\Gamma_+$ and $\Gamma_-$

We now combine what we have learned about the behaviors around the elliptic fixed points (Fig. 10.10) and around the hyperbolic fixed points (Fig. 10.14) which remain after break-up of the rational circles under weak perturbation. Fig. 10.15 attempts to illustrate the situation, but it is only a crude approximation to the full picture. As we have already noted, surrounding the secondary elliptic points are secondary ellipses, some of which are rational. These in turn must break up just as we have discussed for the primary circles, leading to tertiary elliptic and hyperbolic points, and this is repeated over and over on finer and finer scales. In spite of its crudeness, however, this sketch does contain the general features of non-integrable systems which we have already noted in the Hénon-Heiles problem.

Fig. 10.15. Rational circle under perturbation, second step

Finally, we emphasize once again that this rich and complex behavior is the *normal* or generic behavior of a conservative mechanical system with two degrees of freedom.

## Exercises

1. Investigate the surface of section ( $y = 0,\ p_y > 0$, E fixed) for the two-dimensional oscillator with Hamiltonian (refer to exercise 9.02)
$$H = \frac{p_x^2}{2m} + \frac{p_y^2}{2m} + \frac{1}{2}m\omega_x^2 x^2 + \frac{1}{2}m\omega_y^2 y^2.$$
In particular, examine the nature of the sequences of points resulting from various starts,
(a) if $\omega_x/\omega_y$ is an irrational number;
(b) if $\omega_x/\omega_y$ is a rational number, $\omega_x/\omega_y = r/s$.

2. A particle of mass m moves in a (two-dimensional) central force with potential
$$V = -\frac{k}{r} + \frac{h}{r^2}.$$
Using a computer or otherwise, plot the sequences of points $(x, p_x)$ in the surface of section ( $y = 0,\ p_y > 0$, E fixed) which result from representative starts (refer to exercises 1.13 and 9.06).

3. Consider a system with Hamiltonian
$$H = \frac{p^2}{2m} + \frac{1}{2}k_0 x^2 + \frac{1}{2}k_1 x^2.$$
This is, of course, a simple harmonic oscillator with spring constant $k_0 + k_1$ and is exactly soluble. Suppose, however, that we regard the term $\frac{1}{2}k_1 x^2$ as a perturbation. Find, to first order in the perturbation,

(a) the canonical transformation from the unperturbed action-angle variables to the perturbed action-angle variables;
(Ans.: the generator is $G = (k_1 I'/4k_0)\sin 2\phi$)
(b) the Hamiltonian for the perturbed action-angle variables. Hence find the first order correction to the frequency of oscillation. Compare with the exact result.

4. Find the continued fraction expansion of the following numbers, and write down the first five or so continued fraction approximates. Verify that these are closer to the number than $(\text{denominator})^{-2}$. (a) 157/225; (b) $\sqrt{2}$; (c) the **golden ratio**, $\gamma = (\sqrt{5}-1)/2$; (d) the base of natural logarithms, e.

5. A model which illustrates many features of non-linear dynamical systems is the **kicked rotator**. A particle moves on a (frictionless) circular track; every $\tau$ seconds it is given a kick in some fixed direction (Fig. 10.16).

Fig. 10.16. Kicked rotator

If the angle just before a kick is $\theta_n$ and the angular velocity is $\omega_n$, then just after the kick the angle is unchanged but the velocity is changed by $-k\sin\theta_n$ where k is the strength of the kick. Just before the next kick the angle and velocity are thus

$$\theta_{n+1} = \theta_n + \omega_{n+1}\tau$$
$$\omega_{n+1} = \omega_n - k\sin\theta_n \,.$$

These equations can be scaled, $\omega\tau \to \psi$, $k\tau \to \gamma$, to give the form

$$\theta_{n+1} = \theta_n + \psi_{n+1}$$
$$\psi_{n+1} = \psi_n - \gamma\sin\theta_n \,.$$

Clearly, the angles $\theta$ and $\theta + 2\pi$ are equivalent, as are the velocities $\psi$ and $\psi + 2\pi$. We can thus confine our attention to the region $-\pi < \theta < \pi$, $-\pi < \psi < \pi$; whenever $\theta$ or $\psi$ moves outside the region, we translate it back by an appropriate multiple of $2\pi$. This mapping in the $(\theta, \psi)$ phase plane, which describes the dynamical behavior of the kicked rotator and, as it turns out, a number of other systems, is sometimes called the **standard map**.[13] Note that this map is area-preserving, $\partial(\theta_{n+1}, \psi_{n+1})/\partial(\theta_n, \psi_n) = 1$. The "path" followed depends on the start $(\theta_0, \psi_0)$ and on the value of the parameter $\gamma$. Your open-ended assignment is to study the behavior of this system for various starts and parameters. To do so to reasonable depth takes considerably more time than the typical exercise, but it is time well spent.

---

[13]Boris V. Chirikov, "A Universal Instability of Many-Dimensional Oscillator Systems," Phys. Rept. **52**, 263-379 (1979).

First, study the system "experimentally" using a computer. Your pictures will look more symmetric if you use instead of $\psi_n$ an "average velocity" $\overline{\psi}_n = \frac{1}{2}(\psi_n + \psi_{n+1})$ which is more closely associated with the point n than $\psi_n$ (before) or $\psi_{n+1}$ (after). Pick some $\gamma$ and a start and have the computer plot the next $10^2$ or $10^3$ iterations; change the start and repeat until you see how things go over the whole phase space; then pick another $\gamma$. For example, $\gamma = 1$, $\theta_0 = 0$, $\overline{\psi}_0 = -3$ to $+3$ step 0.2 gives Fig. 10.17.

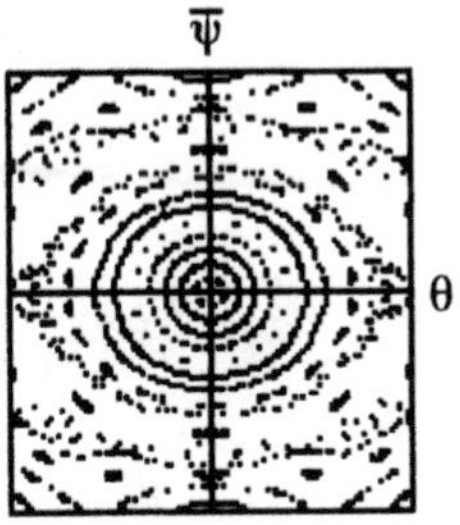


Fig. 10.17. Phase plane for the kicked rotator

You may wish to investigate the sequence of $\gamma$'s at which a sequence of cycles becomes unstable. You will notice that the "distance" $\gamma_{n+1} - \gamma_n$ from one transition to the next tends to zero, but the ratio $\frac{\gamma_n - \gamma_{n-1}}{\gamma_{n+1} - \gamma_n}$ of successive distances tends to some finite number (**Feigenbaum number**). You may also notice that parts of the plane look like scaled down versions of the whole plane and wish to investigate this.

Second, study the system "theoretically" using analytic methods. How much of these patterns can you understand? For example, do you understand the loops near $\theta = \overline{\psi} = 0$ and the conditions under which they exist? the loops near $\theta = 0$, $\overline{\psi} = \pm\pi$? the other loops? Look for fixed points and fixed cycles of points and investigate stability. For example, can you show that (0,0) is a stable fixed point only if $0 < \gamma < 4$? What about the cycles of 2, 3, $\cdots$? Can you picture these physically? What can you say about the irregular behavior which seems to occupy more and more of the phase space as $\gamma$ gets larger?

# INDEX

Printed in the USA
CPSIA information can be obtained
at www.ICGtesting.com
LVHW010710011223
764693LV00003B/227

9 789810 22672